ROWAN UNIVERSITY
CAMPBELL LIBRARY
201 MULLICA HILL RD.
GLASSBORO, NJ 08028-1701

D1253659

HALAL
FOOD PRODUCTION

HALAL
FOOD PRODUCTION

Mian N. Riaz
Muhammad M. Chaudry

CRC PRESS

Boca Raton   London   New York   Washington, D.C.

Library of Congress Cataloging-in-Publication Data

Riaz, Mian N.
  Halal food production / Mian N. Riaz, Muhammad M. Chaudry.
    p. cm.
  Includes bibliographical references and index.
  ISBN 1-58716-029-3 (alk. paper)
    1. Food industry and trade. I. Chaudry, Muhammad M. II. Title.

TP370.R47 2003
297.5'76—dc22

2003055483

This book contains information obtained from authentic and highly regarded sources. Reprinted material is quoted with permission, and sources are indicated. A wide variety of references are listed. Reasonable efforts have been made to publish reliable data and information, but the author and the publisher cannot assume responsibility for the validity of all materials or for the consequences of their use.

Neither this book nor any part may be reproduced or transmitted in any form or by any means, electronic or mechanical, including photocopying, microfilming, and recording, or by any information storage or retrieval system, without prior permission in writing from the publisher.

The consent of CRC Press LLC does not extend to copying for general distribution, for promotion, for creating new works, or for resale. Specific permission must be obtained in writing from CRC Press LLC for such copying.

Direct all inquiries to CRC Press LLC, 2000 N.W. Corporate Blvd., Boca Raton, Florida 33431.

Trademark Notice: Product or corporate names may be trademarks or registered trademarks, and are used only for identification and explanation, without intent to infringe.

Visit the CRC Press Web site at www.crcpress.com

© 2004 by CRC Press LLC

No claim to original U.S. Government works
International Standard Book Number 1-58716-029-3
Library of Congress Card Number 2003055483
Printed in the United States of America 1 2 3 4 5 6 7 8 9 0
Printed on acid-free paper

3 3001 00880 4790

DEDICATION

This book is dedicated to food scientists who continuously strive
to seek knowledge to serve humankind

PREFACE

The word *halal* has become quite common in the Western food industry in the past 2 decades, primarily due to the export of food products to the Middle East and Southeast Asia. The meaning of this Arabic word, "permitted" or "lawful," is very clear. Nevertheless, its practical interpretation varies a great deal among food-importing countries, as does its understanding by companies that produce food. Students of food science and technology generally are not taught about dietary requirements of different religions and ethnic groups and are only exposed to concepts such as kosher, halal, and vegetarian in industry, whereas product development, quality assurance, procurement, and other key personnel are forced to learn about these concepts to meet their customers' requirements.

The books currently in the market for people interested in the subject of halal have been published to help Muslim consumers decide what to eat and what to avoid among the foods already present in the marketplace. There is no book written for the food industry itself that provides the information industry needs to produce food products that meet the needs of both domestic and international consumers.

Both authors, Mian N. Riaz (working at a university) and Muhammad M. Chaudry (working in the food industry), have recognized this gap in the vital information about halal available to the food professionals. Both authors are food scientists, with collectively more than 30 years of practical experience in this area. This book is the result of their practical experience and knowledge in halal food requirements and halal certification.

This book is written to summarize some of the fundamentals to be considered in halal food production. It is an excellent starting point for the food scientist and technologist and other professionals who are in the halal food business. There is a wealth of information about halal food laws and regulations, general guidelines for halal food production, domestic and international halal food markets, and trade and import requirements for different countries. The book also covers specific halal production requirements for meat, poultry, dairy products, fish, seafood, cereal, confectionary, and food supplements. The role of gelatin, enzymes, alcohol, and other questionable ingredients for halal food production is addressed in some detail. Guidelines with examples of labeling, packaging, and coatings for halal food are also presented. The new topic of biotechnology and genetically modified organisms (GMOs) in halal food production is discussed. A brief discussion of the growing concern about animal feed is also provided. A complete chapter is dedicated to the differences among halal, kosher, and vegetarian food production. For food companies that would like their products to be certified halal, a procedure is included for obtaining halal certification. This book also contains 14 appendices, which cover halal food-related information that can be used as guidelines by halal food processors.

The authors believe that this book can serve as a source of information to all who are involved or would like to be involved in any aspect of the halal food business. For persons who are new to this area, this book will serve as a guide for understanding and properly selecting food ingredients for processing halal foods. In view of the growing halal food markets worldwide in food service, branded packaged foods, and direct-marketed products as well as food ingredients, both academia and industry will benefit from this work.

They are deeply indebted to a number of individuals who provided information and inspiration, and guided us in the right direction to complete this book.

ACKNOWLEDGMENT

Completion of this book provides an opportunity to recognize a number of very important individuals. We particularly express our heartfelt gratitude to Dr. Joe Regenstein of Cornell University for his exhaustive suggestions and guidance toward the preparation of the manuscript. His combination of sage advice, the capacity to listen and discuss, the ability to challenge us to expand, and a never-ending willingness to share his time will always be greatly valued. He was extremely helpful in providing the most accurate information and critique throughout this work.

We express our thanks and gratitude to Eleanor Riemer for her help with the final editing and organization, and, more importantly, for motivating us in writing this book. Haider Khattak, Grace Tjahjono, M. Ayub Khan, Roger Othman, Mohamed Sadek, Laura Segreti, and several others provided us very valuable assistance in collecting and compiling the information. Thanks are also due to all friends, family members, and colleagues, who provided help in numerous ways.

Mian M. Riaz
Muhammad M. Chaudry

AUTHORS

Mian N. Riaz, Ph.D., received B.S and M.S. degrees in food technology from the University of Agriculture, Faisalabad, Pakistan, and a Ph.D. in food science from the University of Maine, Orono. He is currently head of the Extrusion Technology Program and a research scientist at the Food Protein Research and Development Center at Texas A&M University, College Station. Dr. Riaz is also on the graduate faculty in the Food Science and Technology Program at Texas A&M University.

Muhammad Munir Chaudry, Ph.D., is currently director of the halal certification program at the Islamic Food and Nutrition Council of America, Chicago, Illinois. Until 1993, he was vice president of total quality and quality assurance at the Heller Seasonings and Ingredients, Inc., Bedford Park, Illinois. Dr. Chaudry received his B.Sc. (Hons) and M.Sc. (Hons) in food technology from the West Pakistan Agricultural University, Lyallpur, Pakistan; an M.S. in food science from the American University of Beirut, Lebanon; and a Ph.D. in food science from the University of Illinois, Urbana-Champaign.

CONTENTS

1

INTRODUCTION

The food industry, like any other industry, responds to the needs and desires of the consumer. People all over the world are now more conscious about foods, health, and nutrition. They are interested in eating healthy foods that are low in calories, cholesterol, fat, and sodium. Many people are interested in foods that are organically produced without the use of synthetic pesticides and other nonnatural chemicals. The ethnic and religious diversity in America and Europe has encouraged the food industry to prepare products which are suitable to different groups such as the Chinese, Japanese, Italian, Indian, Mexican, Seventh Day Adventist, vegetarian, Jewish, and Muslim.

Islam is the world's second largest religion, also the fastest growing, both globally and in the U.S. More than 7 million Muslims live in the U.S. (Cornell University, 2002), and the worldwide Muslim population is ca. 1.3 billion (Chaudry, 2002). The Muslim population is estimated to reach 12.2 million in 2018 in the U.S. (*USA Today*, 1999). Islam is not merely a religion of rituals — it is a way of life. Rules and manners govern the life of the individual Muslim. In Islam, eating is considered a matter of worship of God, like ritual prayers. Muslims follow the Islamic dietary code, and foods that meet that code are called halal (lawful or permitted). Muslims are supposed to make an effort to obtain halal food of good quality. It is their religious obligation to consume only halal food. For non-Muslim consumers, halal foods often are perceived as specially selected and processed to achieve the highest standards of quality.

Between 300 and 400 million Muslims are estimated to live as minorities in different nations of the world, forming a part of many different cultures and societies. In spite of their geographic and ethnic diversity, all Muslims follow their beliefs and the religion of Islam. Halal is a very important and integral part of religious observance for all Muslims. Hence, halal constitutes a universal standard for a Muslim to live by.

1

By definition, halal foods are those that are free from any component that Muslims are prohibited from consuming. According to the Quran (the Muslim scripture), all good and clean foods are halal. Consequently, almost all foods of plant and animal origin are considered halal except those that have been specifically prohibited by the Quran and the Sunnah (the life, actions, and teachings of the Prophet Muhammad).

Until now, there has been no book available combining the religious and production issues that can guide food manufacturers in understanding halal food production. Producing halal food is similar to producing regular foods, except for certain basic requirements, which will be discussed in this book. Halal foods can be processed by using the same equipment and utensils as regular food, with a few exceptions or changes. In the chapters that follow, food manufacturers will learn the requirements of halal food production and gain some knowledge about Muslims and the Muslim markets.

The book is divided into various chapters covering halal laws in general, production guidelines for various product types (including meat and poultry, fish and seafood, dairy products, cereals, food ingredients), labeling, biotechnology, and several other areas of concern to the halal consumers.

The book presents the laws and regulations in a format understandable by non-Muslims. Terminology and concepts that are generally associated with religious jurisprudence have been avoided wherever possible. The laws have been translated into general guidelines for the food industry and kindred product industries. Several chapters have been devoted to specific industries in which the authors feel that halal activity is currently the greatest:

- Meat and poultry products (Chapter 6) — this is the most highly regulated segment of the food industry in regard to halal requirements. Out of five prohibited food categories, four belong to this group.
- Dairy products (Chapter 7) — cheese and whey proteins have received wide acceptance in nondairy food products. Controversy over the use of porcine enzymes before the development of chymosin-type products as rennet replacers or extenders still continues among the Muslim consumers. We have tried to project a balanced picture of these products' requirements, with special emphasis on enzymes.
- Fish and seafood (Chapter 8) — although not very significant in international trade, fish and seafood products are subject to more controversy than any other food group among Muslim consumers. This chapter covers explanations of the status of various types of fish, shellfish, crustaceans, and other seafood products.

- Cereal-based products (Chapter 9) — cereal-based products, candy, and other products have been discussed in much briefer formats because of the relatively fewer controversial issues in these products for different Muslim consumers. Information on nutritional supplements (Chapter 10) has been included to reflect the high visibility and demand for halal certified products throughout the world, but specifically in the Southeast Asian countries.
- Food ingredients (Chapter 14) — this chapter covers the many diverse items used all across the food industry and produced from plants, animals, microorganisms, or by synthetic processes. More emphasis has been placed on flavors, amino acids, oils and extracts, and blended products. Two key food ingredients that require more extensive discussion are included as separate chapters — gelatin (Chapter 11), enzymes (Chapter 12), and alcohol (Chapter 13).

In the chapters covering halal requirements for different products categories, the concept of hazard analysis and critical control points (HACCP) has been used to identify halal control points (HCPs). The objective here is not to replace HACCP, which address the food safety issues, but to complement these requirements by adding key points for halal compliance. HCPs have been presented in an easy-to-understand flowchart format. By using the guidelines provided herein, the companies are encouraged to devise their own HCPs and include them in their standard operating procedures as a self-compliance tool.

Marketing and trade aspects of halal foods have been included in two chapters: one covers the domestic and international trade of halal food products, and the second covers import requirements for various Muslim countries as well as the Muslim population and level of activities relevant to halal.

Finally, information has been included about procedures for getting halal certification. Food manufacturers can obtain supervision from different halal food certifying agencies such as the Islamic Food and Nutrition Council of America (IFANCA) as well as reliable information about Islam and Muslims in North America and their critical food issues. These halal-certifying agencies provide consultation services and help food industry professionals develop products that comply with Islamic food laws. These agencies also offer supervision and certification for halal foods, consumer products, and halal-slaughtered meat and poultry. Their registered trademark certification symbol, for example, the Crescent M®, appears on many product packages. The demand for halal products and number of Muslim consumers can easily be an inducement for manufacturers to provide halal products. Halal markings are an important part of the general acceptance of halal products by the Muslim consumer worldwide. Certain key information with respect to halal food production is included in several appendices, such as the relevant section of Codex Alimentarius and the guidelines for labeling halal products in a number of representative countries, the halal status of common and E-numbered ingredients, halal industrial production standards, and halal food laws of several states in the U.S.

REFERENCES

Chaudry, M.M. 2002. Halal certification process, paper presented at *Market Outlook: 2002 Conference, Toward Efficient Egyptian Processed Food Export Industry in a Global Environment,* Cairo, Egypt.

Cornell University. April 2002. Study on American Muslim, survey sponsored by Bridges TV, Orchard Park, NY.

Facts, figures about Islam, Muslim future: projected Muslim population in the USA, *USA Today,* June 25, 1999, p. 12B.

2

HALAL FOOD LAWS AND REGULATIONS

The basic guidance about the halal food laws is revealed in the Quran (the divine book) from God (the Creator) to Muhammad (the Prophet) for all people. The food laws are explained and put into practice through the Sunnah (the life, actions, and teachings of Muhammad) as recorded in the Hadith (the compilation of the traditions of Muhammad).

In general, everything is permitted for human use and benefit. Nothing is forbidden except what is prohibited either by a verse of the Quran or an authentic and explicit Sunnah of Muhammad. These rules of Shariah (Islamic law) bring freedom for people to eat and drink anything they like as long as it is not haram (prohibited).

There are five fundamental pillars of belief in Islam: (1) to believe that there is no god but Allah, and Muhammad is his last prophet; (2) to pray five times in a day; (3) to give zakat (charity) to the poor; (4) to fast in the month of Ramadan; and (5) to perform the pilgrimage to Mecca once in a lifetime (if one can afford it). In addition, guidelines direct the daily life of a Muslim. Included in these guidelines is a set of dietary laws intended to advance wellness. These laws are binding on the faithful and must be observed at all times, even during pregnancy, periods of illness, or traveling (Twaigery and Spillman, 1989). The life of a Muslim revolves around the concept of halal and haram. The laws are quite comprehensive, because they are applicable not only to eating and drinking, but also to earning one's living, dress code, and dealing with others. This discussion will focus primarily on food.

Food is considered one of the most important factors for interaction among various ethnic, social, and religious groups. All people are concerned about the food they eat: Muslims want to ensure that their food is halal; Jews that their food is kosher; Hindus, Buddhists, and certain other groups that their food is vegetarian. Muslims follow clear guidelines in the selection of their food. The principles behind halal food are described here.

PRINCIPLES REGARDING PERMISSIBILITY OF FOODS

Eleven generally accepted principles pertaining to halal (permitted) and haram (prohibited) in Islam provide guidance to Muslims in their customary practices (Al-Qaradawi, 1984):

- The basic principle is that all things created by God are permitted, with a few exceptions that are specifically prohibited.
- To make lawful and unlawful is the right of God alone. No human being, no matter how pious or powerful, may take this right into his own hands.
- Prohibiting what is permitted and permitting what is prohibited is similar to ascribing partners to God.
- The basic reasons for the prohibition of things are impurity and harmfulness. A Muslim is not required to know exactly why or how something is unclean or harmful in what God has prohibited. There might be obvious reasons, and there might be obscure reasons.
- What is permitted is sufficient, and what is prohibited is then superfluous. God prohibited only things that are unnecessary or dispensable while providing better alternatives.
- Whatever is conducive to the "prohibited" is in itself prohibited. If something is prohibited, anything leading to it is also prohibited.
- Falsely representing unlawful as lawful is prohibited. It is unlawful to legalize God's prohibitions by flimsy excuses. To represent lawful as unlawful is also prohibited.
- Good intentions do not make the unlawful acceptable. Whenever any permissible action of the believer is accompanied by a good intention, his action becomes an act of worship. In the case of haram, it remains haram no matter how good the intention, how honorable the purpose, or how lofty the goal. Islam does not endorse employing a haram means to achieve a praiseworthy end. Indeed, it insists not only that the goal be honorable, but also that the means chosen to attain it be proper. "The end justifies the means" and "Secure your right even through wrongdoing" are maxims not acceptable in Islam. Islamic law demands that the right should be secured through just means only.
- Doubtful things should be avoided. There is a gray area between clearly lawful and clearly unlawful. This is the area of "what is

doubtful." Islam considers it an act of piety for Muslims to avoid doubtful things, and for them to stay clear of the unlawful. Muhammad said (Sakr, 1994): "Halal is clear and haram is clear; in between these two are certain things that are suspected. Many people may not know whether these items are halal or haram. Whosoever, leaves them, he is innocent toward his religion and his conscience. He is, therefore, safe. Anyone who gets involved in any of these suspected items, he may fall into the unlawful and the prohibition. This case is similar to the one who wishes to raise his animals next to a restricted area, he may step into it. Indeed the restrictions of Allah are the unlawful."

■ Unlawful things are prohibited to everyone alike. Islamic laws are universally applicable to all races, creeds, and sexes. There is no favored treatment of any privileged class. Actually, in Islam, there are no privileged classes; hence, the question of preferential treatment does not arise. This principle applies not only among Muslims but between Muslims and non-Muslims as well.

■ Necessity dictates exceptions. The range of prohibited things in Islam is very narrow, but emphasis on observing the prohibitions is very strong. At the same time, Islam is not oblivious to the exigencies of life, to their magnitude, or to human weakness and capacity to face them. It permits the Muslim, under the compulsion of necessity, to eat a prohibited food in quantities sufficient to remove the necessity and thereby survive.

Five major terms are used to describe the permissibility of food:

■ Halal means permissible and lawful. It applies not only to meat and poultry, but also to other food products, cosmetics, and personal care products. The term also applies to personal behavior and interaction with the community.

■ Haram means prohibited. It is directly opposite of halal.

■ Mashbooh is something questionable or doubtful, either due to the differences in scholars' opinions or the presence of undetermined ingredients in a food product.

■ Makrooh is a term generally associated with someone's dislike for a food product or, while not clearly haram, is considered dislikeable by some Muslims.

■ Zabiha or dhabiha is a term often used by Muslims in the U.S to differentiate meat that has been slaughtered by Muslims as opposed to being slaughtered by Ahlul Kitab (Jews or Christians) or without religious connotation.

HALAL AND HARAM

General Quranic guidance dictates that all foods are halal except those that are specifically mentioned as haram. All foods are made lawful according to the Muslim scripture *The Glorious Quran* [Arabic text and English rendering by Pickthall (1994)]:

> *O ye who believe! Eat of the good things wherewith We have provided you, and render thanks to Allah, if it is (indeed) He whom ye worship.*

> *Chapter II, Verse 172*

The unlawful foods are specifically mentioned in *The Glorious Quran* in the following verses:

> *He hath forbidden you only carrion, and blood, and swine flesh, and that which hath been immolated to (the name of) any other than Allah...*

> *Chapter II, Verse 173*

> *Forbidden unto you (for food) are carrion and blood and swine flesh, and that which hath been dedicated unto any other than Allah, and the strangled, and the dead through beating, and the dead through falling from a height, and that which hath been killed by (the goring of) horns, and the devoured of wild beasts saving that which ye make lawful (by the death-stroke) and that which hath been immolated unto idols. And (forbidden is it) that ye swear by the divining arrows. This is an abomination...*

> *Chapter V, Verse 3*

Consumption of alcohol and other intoxicants is prohibited according to the following verse:

> *O ye who believe! Strong drink and games of chance, and idols and divining arrows are only an infamy of Satan's handiwork. Leave it aside in order that ye may succeed.*

> *Chapter V, Verse 90*

Meat is the most strictly regulated of the food groups. Not only are blood, pork, and the meat of dead animals or those immolated to other than God strongly prohibited, it is also required that halal animals be slaughtered while pronouncing the name of God at the time of slaughter.

Eat of that over which the name of Allah hath been mentioned, if ye are believers in His revelations.

Chapter VI, Verse 118

And eat not of that whereon Allah's name hath not been mentioned, for lo! It is abomination. Lo! the devils do inspire their minions to dispute with you. But if ye obey them, ye will be in truth idolaters.

Chapter VI, Verse 121

Accordingly, all foods pure and clean are permitted for consumption by the Muslims except the following categories, including any products derived from them or contaminated with them:

■ Carrion or dead animals
■ Flowing or congealed blood
■ Swine, including all by-products
■ Animals slaughtered without pronouncing the name of God on them
■ Animals killed in a manner that prevents their blood from being fully drained from their bodies
■ Animals slaughtered while pronouncing a name other than God
■ Intoxicants of all types, including alcohol and drugs
■ Carnivorous animals with fangs, such as lions, dogs, wolves, or tigers
■ Birds with sharp claws (birds of prey), such as falcons, eagles, owls, or vultures
■ Land animals such as frogs or snakes

From the Quranic verses, the hadith, and their explanations and commentary by Muslim scholars, the Islamic food (dietary) laws are deduced. Additional verses in *The Glorious Quran* related to food and drinks are as follows:

O' mankind! Eat of that which is lawful and wholesome in the earth, and follow not the footsteps of the devil. Lo! he is an open enemy for you.

Chapter II, Verse 168

Oh ye who believe! Fulfill your indentures. The beast of cattle is made lawful unto you (for food) except...

Chapter V, Verse 1

They ask thee (O Muhammad) what is made lawful for them. Say: (all) good things are made lawful for you. And those beasts and birds of prey which ye have trained as hounds are trained, ye teach them that which Allah taught you; so eat of that which they catch for you and mention Allah's name upon it, and observe your duty to Allah. Allah is swift to take account.

Chapter V, Verse 4

This day are (all) good things made lawful for you. The food of those who have received Scripture is lawful for you and your food is lawful for them...

Chapter V, Verse 5

Oh ye who believe! Forbid not the good things, which Allah had made lawful for you, and transgress not. Lo Allah loveth not transgressors.

Chapter V, Verse 87

Eat of that which Allah hath bestowed on you as food lawful and good, and keep your duty to Allah in whom ye are believers.

Chapter V, Verse 88

How should ye not eat of that over which the name of Allah hath been mentioned, when He hath explained unto you that which is forbidden into you, unless ye are compelled thereto.

Chapter VI, Verse 119

And eat not of that whereon Allah's name hath not been mentioned, for lo! It is abomination. Lo! the devils do inspire their minions to dispute with you. But if ye obey them, ye will be in truth idolaters.

Chapter VI, Verse 121

And of the cattle He produceth production some for burden and some for food; Eat of that which Allah hath bestowed upon you, and follow not the footsteps of the devil, for lo! he is an open foe to you.

Chapter VI, Verse 142

Say: "I find not in that which is revealed unto me ought prohibited to an eater that he eat thereof except it be carrion, or blood poured forth, or swineflesh — for that verily is foul — or the abomination which was immolated to the name of other than Allah. But who so is compelled (there to), neither craving nor transgressing, (for him) Lo! Your Lord is forgiving, merciful.

Chapter VI, Verse 145

So eat of the lawful and good food, which Allah has provided for you and thank the bounty of your Lord if it is Him ye serve.

Chapter XVI, Verse 114

He hath forbidden for you only carrion, and blood and the swine flesh, and that which hath been immolated in the name of any other than Allah; but he who is driven thereto, neither craving nor transgressing, Lo! then Allah is forgiving, merciful.

Chapter XVI, Verse 115

The haram foods are mainly pork, alcohol, blood, dead animals, and animals slaughtered while reciting a name other than that of God. This may also include halal items that have been contaminated or mixed with haram items. In general, most Muslims deem meat and poultry items not slaughtered in the name of God to be haram or makrooh at best.

BASIS FOR THE PROHIBITIONS

In the Islamic faith, Allah is the Almighty God. He has no partners. The first requirement of a Muslim is to declare: "There is no god but God (Allah)." So everything has to be dedicated to God only. There is no challenge to this fact, and no explanations are required or necessary. The basis for the prohibition of the above categories is purely and strictly Quranic guidance. However, some scientists have attempted to explain or justify some of these prohibitions based on their scientific understanding as follows:

- Carrion and dead animals are unfit for human consumption because the decaying process leads to the formation of chemicals which are harmful to humans (Awan, 1988).
- Blood that is drained from the body contains harmful bacteria, products of metabolism, and toxins (Awan, 1988; Hussaini and Sakr, 1983).
- Swine serves as a vector for pathogenic worms to enter the human body. Infection by *Trichinella spiralis* and *Taenia solium* are not uncommon. Fatty acid composition of pork fat has been mentioned as incompatible with the human fat and biochemical systems (Awan, 1988; Hussaini and Sakr, 1983; Sakr, 1993).
- Intoxicants are considered harmful for the nervous system, affecting the senses and human judgment. In many cases they lead to social and family problems and even loss of lives (Al-Qaradawi, 1984; Awan, 1988).

Although these explanations may or may not be sound, the underlying principle behind the prohibitions remains the divine order, which appears in the Quran in several places: "Forbidden unto you are…" is what guides a Muslim believer.

HOW DOES ONE TRANSLATE MAJOR PROHIBITIONS INTO PRACTICE IN TODAY'S INDUSTRIAL ENVIRONMENT?

Let us look at how the laws are translated into practice:

- Carrion and dead animals — it is generally recognized that eating carrion is offensive to human dignity, and probably nobody consumes it in the modern civilized society. However, there is a chance of an animal dying from the shock of stunning before it is properly slaughtered. This is more common in Europe than in North America. The meat of such dead animals is not proper for Muslim consumption (Chaudry, 1992).
- Proper slaughtering — there are strict requirements for the slaughtering of animals: the animal must be of a halal species, that is, cattle, lamb, etc.; the animal must be slaughtered by a Muslim of proper age; the name of God must be pronounced at the time of slaughter; and the slaughter must be done by cutting the throat of the animal in a manner that induces rapid, complete bleeding and results in the quickest death.

Certain other conditions should also be observed. These include considerate treatment of the animal, giving it water to prevent thirst, and using a sharp knife. These conditions ensure the humane treatment of animals before and during slaughter. Any by-products or derived ingredients must also be from duly slaughtered animals to be good for Muslim consumption.

- Swine — pork, lard, and their by-products or derived ingredients are categorically prohibited for Muslim consumption. All chances of cross-contamination from pork into halal products must be thoroughly prevented. In fact, in Islam, the prohibition extends beyond eating. For example, a Muslim must not buy, sell, raise, transport, slaughter, or in any way directly derive benefit from swine or other haram media.
- Blood — blood that pours forth (liquid blood) is generally not offered in the marketplace or consumed, but products made from blood and ingredients derived from it are available. There is general agreement among religious scholars that anything made from blood is unlawful for Muslims.
- Alcohol and other intoxicants — alcoholic beverages such as wine, beer, and hard liquors are strictly prohibited. Foods containing added amounts of alcoholic beverages are also prohibited because such foods, by definition, become impure. Nonmedical drugs and other intoxicants that affect a person's mind, health, and overall performance are prohibited too. Consuming these directly or incorporated into foods is not permitted. However, there are certain acceptable allowances for naturally present alcohol or alcohol used in processing of food, as discussed in Chapter 13.

Foods are broadly categorized into four groups for the ease of establishing their halal status and formulating guidelines for the industry.

- Meat and poultry — this group contains four out of five haram (prohibited) categories. Hence, higher restrictions are observed here. Animals must be halal. One cannot slaughter a pig the Islamic way and call it halal. Animals must be slaughtered by a sane Muslim while pronouncing the name of God. A sharp knife must be used to severe the jugular veins, carotid arteries, trachea, and esophagus, and blood must be drained out completely. Islam places great emphasis on humane treatment of animals, so dismemberment must not take place before the animal is completely dead, as described earlier.
- Fish and seafood — to determine the acceptability of fish and seafood, one has to understand the rules of different schools of Islamic jurisprudence, as well as the cultural practices of Muslims living in different regions. All Muslims accept fish with scales; however, some

groups do not accept fish without scales such as catfish. There are even greater differences among Muslims about seafood, such as molluscs and crustaceans. One must understand the requirements in various regions of the world, for example, for exporting products containing seafood flavors.

■ Milk and eggs — from the halal animals are also halal. The predominant source of milk in the West is the cow, and the predominant source of eggs is the chicken. All other sources are required to be labeled accordingly. There are a variety of products made from milk and eggs. Milk is used for making cheese, butter, and cream. Most of the cheeses are made with various enzymes, which could be halal if made with microorganisms or halal-slaughtered animals. The enzymes could be haram if extracted from porcine sources or questionable when obtained from non-halal-slaughtered animals. Similarly, emulsifiers, mold inhibitors, and other functional ingredients from nonspecified sources can make milk and egg products doubtful to consume.

■ Plants and vegetables — these materials are generally halal except alcoholic drinks or other intoxicants. However, in modern-day processing plants, vegetables and meats might be processed in the same plant and on the same equipment, increasing the chance of cross-contamination. Certain functional ingredients from animal sources might also be used in the processing of vegetables, which make the products doubtful. Hence, processing aids and production methods have to be carefully monitored to maintain the halal status of foods of plant origin.

From this discussion on laws and regulations it is clear that several factors determine the halal or haram status of a particular foodstuff. It depends on its nature, how it is processed, and how it is obtained. As an example, any product from pig would be considered as haram because the material itself is haram. Similarly, beef from an animal that has not been slaughtered according to Islamic rites would still be considered unacceptable. And, of course, a stolen foodstuff or any products that are acquired through means that are incompatible with Islamic teaching would also be haram. Food and drink that are poisonous or intoxicating are obviously haram even in small quantities because they are harmful to health.

REFERENCES

Al-Qaradawi, Y. 1984. *The Lawful and Prohibited in Islam,* The Holy Quran Publishing House, Beirut, Lebanon.

Awan, J.A. 1988. Islamic food laws. I. Philosophy of the prohibition of unlawful foods, *Sci. Technol. Islam. World,* 6(3), 151.

Chaudry, M.M. 1992. Islamic food laws: philosophical basis and practical implications, *Food Technol.*, 46(10), 92, 93, 104.

Hussaini, M.M. and Sakr, A.H. 1983. *Islamic Dietary Laws and Practices*, Islamic Food and Nutrition Council of America, Bedford Park, IL.

Pickthall, M.M. 1994. Arabic text and English rendering of *The Glorious Quran*, Library of Islam, Kazi Publications, Chicago, IL.

Sakr, A.H. 1993. Current issues of Halal foods in North America, *Light,* May/June, 3(3), 22-24.

Sakr, A.H. 1994. Halal and Haram defined. In: *Understanding Halal Foods-Fallacies and Facts*. Foundation for Islamic Knowledge, Lombard, IL.

Twaigery, S. and Spillman, D. 1989. An introduction to Moslem dietary laws, *Food Technol.*, 7, 88-90.

3

GENERAL GUIDELINES FOR HALAL FOOD PRODUCTION

In Chapter 2, we discussed halal laws and regulations, and in this chapter we will try to explain how these laws and regulations apply to real situations in the production of halal food. The guidelines in this chapter are general in nature, and specific guidelines for different product types appear in subsequent chapters. Here, foods are broadly classified into four groups to establish their halal status and to formulate guidelines for halal production and certification.

MEAT AND POULTRY

It is understood that meat of only halal animals is allowed for consumption by Muslims. An animal must be of halal species to be slaughtered as halal. The animal must be slaughtered by a sane adult Muslim while pronouncing the name of God. A sharp knife must be used to cut the throat in a manner that induces thorough removal of blood and quick death. Islam places great emphasis on humane treatment of animals. The animals must be raised, transported, handled, and held under humane conditions. However, these are only desirable actions and mishandling of animals does not make their meat haram. Stunning of animals before nonreligious slaughtering is generally accepted in the U.S. and Canada where methods of stunning generally are non-lethal. In many European countries, the type and severity of stunning usually kills the animals before bleeding, which makes it unacceptable for halal. Moreover, dismemberment (i.e., cutting off the horns, ears, lower legs) of an animal must not take place before the animal is completely dead (Chaudry, 1997).

Conditions and Method of Slaughtering (Dhabh or Zabh*)

Dhabh is a clearly defined method of killing an animal for the sole purpose of making its meat fit for human consumption. The word dhabh in Arabic means purification or rendering something good or wholesome. The dhabh method is also called dhakaat in Arabic, which means purification or making something complete.

The following conditions must be fulfilled for dhabh to meet the requirements of the shariah (jurisprudence).

The Slaughter Person

The person performing the act of dhabh must be of sound mind and an adult Muslim. The person can be of either sex. If a person lacks or loses the competence through intoxication or loss of mental abilities, he or she may not perform halal slaughter. The meat of an animal killed by an idolater, a nonbeliever, or someone who has apostatized from Islam is not acceptable.

The Instrument

The knife used to perform dhabh must be extremely sharp to facilitate quick cutting of the skin and severing of blood vessels to enable the blood to flow immediately and quickly, in other words, to bring about an immediate and massive hemorrhage. Muhammad said: "Verily God has prescribed proficiency in all things. Thus if you kill, kill well; and if you perform dhabh, perform it well. Let each of you sharpen his blade and let him spare suffering to the animal he slays" (Khan, 1991). Muhammad is reported to have forbidden the use of an instrument that killed the animal by cutting its skin but not severing the jugular vein. It is also a tradition not to sharpen the knife in front of the animal about to be slaughtered.

The Cut

The incision should be made in the neck at some point just below the glottis and the base of the neck. Traditionally, camels used to be slayed by making an incision anywhere on the neck. This process is called nahr, which means spearing the hollow of the neck. With modern restraining methods and stunning techniques, this procedure might not be appropriate any longer. The trachea and the esophagus must be cut in addition to

---

* Dhabh or zabh is the same word pronounced differently. It means the method and conditions of slaughtering.

the jugular veins and the carotid arteries. The spinal cord must not be cut. The head is therefore not to be severed completely. It is interesting to note that the kosher kill is very similar to the traditional method of dhabh described, except that the invocation is not made on each animal.

The Invocation

Tasmiyyah or invocation means pronouncing the name of God by saying Bismillah (in the name of Allah) or Bismillah Allahu Akbar (in the name of God, God is Great) before cutting the neck. Opinions differ somewhat on the issue of invocation, according to three of the earliest jurists. According to Imam Malik, if the name of God is not mentioned over the animal before slaughtering, the meat of such animal is haram or forbidden, whether one neglects to say Bismillah intentionally or unintentionally. According to the jurist Abu Hanifah, if one neglects to say Bismillah intentionally, the meat is haram; if the omission is unintentional, the meat is halal. According to Imam Shaf'ii, whether one neglects to say Bismillah intentionally or unintentionally before slaughtering, the meat is halal so long as the person is competent to perform dhabh (Khan, 1991).

It is also enough to state here that the above tradition does not prove that the pronouncing of God's name is not obligatory in performing dhabh. In fact, the tradition emphasizes that the pronouncing of God's name was a widely known matter and was considered an essential condition of dhabh (Khan, 1991).

Abominable Acts in Slaying of Animal

- It is abominable to first throw the animal down on its side and sharpen the knife afterwards. It is narrated that the Prophet once passed by a person who, having cast a goat to the ground, was pressing its head with his foot and sharpening his knife while the animal was watching. The Prophet said, "Will this goat not die before being slain? Do you wish to kill it twice? Do not kill one animal in the presence of another, or sharpen your knives before them" (Khan, 1991).
- It is abominable to let the knife reach the spinal cord or to cut off the head of the animal. In South Asia, the term used for cutting of the head, usually by hitting the animal from behind the neck, is called jhatka or killing with a blow. There is general abhorrence in the Muslim community to such killing.
- It is abominable to break the neck of an animal or begin skinning it or cut any parts while it is convulsing or before its life is completely departed. Muhammad said, "Do not deal hastily with the souls (of

animals) before their life departs" (Khan, 1991). It is sometimes the practice in fast-pace commercial slaughterhouses to start removing the horns, ears, and front legs while the animal still seems to be alive. This is against the principles and requirements of dhabh and must be avoided.

■ It is abominable to perform dhabh with a dull instrument. Muhammad commanded that knives be sharpened and be concealed from animals to be slain.

■ It is also abominable to slaughter one animal while the next in line is watching the animal being killed. This is against the humaneness of the process of slaughtering.

From the foregoing description, it is clear that both intention and a precise method are conditions for the validity of dhabh. The insistence on pronouncing the name of God before slaying an animal is meant to emphasize the sanctity of life and the fact that all life belongs to God. Pronouncing the tasmiyyah induces feelings of tenderness and compassion and serves to prevent cruelty. It also reinforces the notion that an animal is being slaughtered in the name of God for food and not for recreational purposes. It is not allowed in Islam to kill an animal for the sole purpose of receiving pleasure out of killing it.

Advantages of Halal Slaughtering

The actual method of dhabh has many advantages. To begin with, the speed of the incision made with the recommended sharp knife shortens the total time to slaughter and seems to inflict less pain than stunning. In a modern slaughterhouse, where animals are stunned before slaying, some of the animals do not become unconscious with one blow and have to be hit more than once.

The method of dhabh allows rapid and efficient bleeding of the animal. It is also obvious that blood being enclosed in a closed circuit can be removed faster by cutting the blood vessels. The force of the beating of the heart puts the blood into circulation. Therefore, the stronger the heart beat, the greater the quantity of blood poured out. It seems that the blood gushes out with dhabh slaughter; while it pours out somewhat slower when the animal has been stunned. The body of the dhabh animal convulses involuntarily more than the stunned animal. Convulsions produce the squeezing or wringing action of the muscles of the body on the blood vessels, which helps to get rid of maximum amount of blood from the meat tissue into the circulation (Khan, 1991).

The physiological conditions described have a bearing on the removal of blood from an animal's body, but they operate fully only if the animal is bled while alive by cutting across its throat and sparing the vertebral column without stunning the brain of the animal in any way (Khan, 1991). With the type of stunning and the force of the blow or shock used in North America, animals are usually alive for several minutes after stunning. The throat is generally cut within the first two minutes after stunning. Due to these reasons, stunning of cattle with captive bolt and poultry with electrified water is practiced in some dhabh slaughter. In some other countries, the blow of stunning is severe enough to kill the animal. In Australia, some organizations contend that stunning renders the animals dead; hence, these organizations do not permit stunning for halal slaughter (AFIC, 2003).

FISH AND SEAFOOD

To determine the acceptability of fish and seafood, one must understand the rules in different schools of Islamic jurisprudence as well as the cultural practices of Muslims living in different regions. Fish with scales are accepted by all denominations and groups of Muslims. Some groups do not consume fish without scales (e.g., catfish). There are additional differences among Muslims about seafood, especially molluscs (e.g., clams, oysters, and squid) and crustaceans (e.g., shrimp, lobster, and crab). The requirements and restriction apply not only to fish and seafood but also to flavors as well as ingredients derived from such products.

MILK AND EGGS

Milk and eggs from halal animals are also halal. Predominantly, milk in the West comes from cows and eggs come from hens. All other sources are required to be labeled accordingly. Numerous products are made from milk and eggs. Milk is used to make cheese, butter, and cream. A variety of enzymes are used in the production of cheeses. Types of enzymes used in the making of cheeses are very important. Enzymes can be halal or haram, depending on their source of origin. Enzymes from microbial sources or halal-slaughtered animals are halal. However, an enzyme from a porcine source is haram. Depending on the enzymes used in production of cheeses or other dairy products, the products are classified as halal, haram, or questionable. On the same basis, other functional additives such as emulsifiers or mold inhibitors should also be screened to take the doubt out of milk or egg products (Riaz, 2000).

PLANT AND VEGETABLE MATERIALS

Foods from plants are halal, with the exception of khamr (intoxicating drinks). In modern processing plants, however, animal or vegetable products might be processed in the same plant on the same equipment, increasing the chances of contamination. For example, in some factories, pork and beans as well as corn are canned on the same equipment. When proper cleaning procedures are used and the halal production segregated from non-halal, contamination can be avoided. Functional ingredients from animal sources, such as antifoams, must also be avoided in the processing of vegetables. This intentional inclusion of haram ingredients into plant and vegetable products may render them as haram. It is evident that processing aids and production methods have to be carefully monitored to maintain halal status of vegetable products.

FOOD INGREDIENTS

Food ingredients are one of the main subjects of concern. Vegetable products, as mentioned earlier, are halal unless they have been contaminated with haram ingredients or contain intoxicating substances. We have already discussed the requirements for animal slaughter and types of seafood permitted for consumption. Here we discuss some of the commonly used ingredients such as gelatin, glycerin, emulsifiers, enzymes, alcohol, animal fat and protein, and flavors and flavorings. Because most of the products fall into questionable or doubtful categories, they require that the majority of manufacturers have their plants inspected and products certified as halal.

Gelatin

The use of gelatin is very common in many food products. Gelatin can be halal if from dhabh-slaughtered animals, doubtful if from animals not slaughtered in a halal manner, or haram if from prohibited animal sources. The source of gelatin is not required to be identified by the Food and Drug Administration (FDA) on product labels. When the source is not known, it can be from either halal or haram sources, hence questionable. Muslims avoid products containing gelatin unless they are certified halal. Common sources of gelatin are pigskin, cattle hides, cattle bones, and, to a smaller extent, fish skins. For formulating, halal products use gelatin from cattle that have been slaughtered in an Islamic manner or from fish.

Glycerin

Glycerin is another ingredient widely used in the food industry. Products containing glycerin are avoided by Muslims because it could be from animal sources. Currently, glycerin from palm oil and other vegetable oils is available for use in halal products.

Emulsifiers

Emulsifiers such as monoglycerides, diglycerides, polysorbates, diacetyl tartaric esters of mono- and di-glycerides (DATEM), and other similar chemicals are another commonly used group of ingredients that can come from halal or haram sources. Some of the companies have started to list the source, especially if it is vegetable, on the labels. If an emulsifier from vegetable sources is used, it is advantageous to indicate that on the label. Emulsifiers from vegetable sources and halal-slaughtered animal sources are halal.

Enzymes

Enzymes are used in many food processes. The most common are the ones used in the cheese and the starch industries. Until a few years ago, the majority of the enzymes used in the food industry were from animal sources; now there are microbial alternatives.

Products such as cheeses, whey powders, lactose, whey protein concentrates, and isolates made from microbial enzymes are halal as long as all other halal requirements are met. Some products made with mixed or animal-based enzymes are haram if porcine enzymes are used; otherwise, they fall in the doubtful category. Bovine rennet and other enzymes from non-halal-slaughtered animals have been accepted by some countries. As more and more microbial enzymes become available, such acceptance will decrease. Use of dairy ingredients in all types of food products is very common, because whey and whey derivatives are an economical source of protein. For the products to be certified halal, dairy ingredients as well as other ingredients must be halal.

Alcohol

Muslims are prohibited from consuming alcoholic beverages, even in small quantities. Alcoholic drinks such as wine and beer should not be added to other products for flavoring or during cooking. Even a small amount of an alcoholic beverage added to a halal product makes it haram (Riaz,

1997). Cooking with wine, beer, and other alcoholic beverages is quite common in the West as well as in China. In Chinese cooking, rice wine is a common ingredient in many recipes. Product formulators and chefs should avoid the use of alcohol in preparing halal products.

Alcohol is so ubiquitous in all biological systems that even fresh fruits contain traces of alcohol. During extraction of essences from fruits, alcohol might get concentrated into the essences. Because such alcohol is naturally present and unavoidable, it does not nullify the halal status of food products in which such essences are present. Furthermore, alcohol in its pure form is used for extracting, dissolving, and precipitating functions in the food industry. As it is the best solvent or chemical available in many cases to carry out certain processes, religious scholars have realized its importance for use in the industry. Ingredients made with alcohol or extracted by using alcohol have become acceptable as long as alcohol is evaporated from the final ingredient. Food ingredients with 0.5% residual or technical alcohol are generally acceptable. However, for consumer items, acceptable limits vary from country to country and one group to another. The Islamic Food and Nutrition Council of America accepts a level of 0.1%, a 1:1000 dilution of alcohol, which is considered an impurity. In halal food laws, if an impurity is not detectable by taste, smell, or sight, it does not nullify the halal status of a food.

A recent article titled *Change of State — Istihala* (Al-Quaderi, 2001) supports this position in the following words. Wine is haram; however, if the same wine turns to vinegar it becomes halal. The use of the vinegar derived from wine is halal as long as no wine remains in it. From these examples, it becomes clear that if an unlawful food item changes state, then the original ruling also changes.

Animal Fat and Protein

Meat and poultry products are not only consumed as staple food items, but are also converted into further processed ingredients to be used in formulating a myriad of nonmeat food products. In the food industry in the U.S. as well as in other industrialized countries, every part of the animal is used in one way or another. Less desirable parts of the carcass and by-products are turned into powders and derived food ingredients, and used as flavoring agents for soups, snacks, etc. Animal fat is purified and converted into animal shortening, emulsifiers, as well as other functional food ingredients. Feathers and hair can be converted into amino acids. Such ingredients would be halal only if the animals are halal and all precautions are taken to eliminate cross-contamination.

Flavors and Flavorings

Flavors and flavorings can be as simple as a single spice such as pepper or as complex as cola flavor or pastrami flavor containing several ingredients. Some of the more complex flavorings can contain over one hundred ingredients of various origins. Thousands of ingredients can be used to create a flavor. These ingredients can be from microorganisms, plants, minerals, petroleum, or animals as well as synthetic sources. For formulating halal food products, the manufacturer has to make sure that any flavors, proprietary mixes, or secret formulas are halal and free from doubtful materials.

SANITATION

During the manufacture of halal products, it is imperative that all possible sources of contamination be eliminated. This can be accomplished through proper scheduling of products as well as by thoroughly cleaning and sanitizing production lines and equipment. For nonmeat products, it is adequate to clean equipment and determine cleanliness by visual observation. A company might treat haram ingredients similar to allergens and make it part of an allergen control program. Chemicals used for cleaning (especially soaps and foams) should be screened to avoid animal fat origin.

SPECIFIC HALAL GUIDELINES

General guidelines for halal vary somewhat from country to country. The following four documents included in the appendices will be helpful to manufacturers of food products for specific markets:

- Codex Alimentarius Guidelines for Use of the Term Halal (Appendix A)
- Halal Industrial Production Standards (Appendix B)
- Malaysian General Guidelines on the Slaughtering of Animals and the Preparation and Handling of Halal Food (Appendix C)
- Singapore's Halal Regulations and Import Requirements (Appendix D)

STATE HALAL LAWS

In 2000, the state of New Jersey passed the Halal Food Protection Act. Minnesota and Illinois followed suit in 2001, enacting their own laws to regulate the halal food industry. These laws address the problem of fraudulent use of the term halal or other such terms to produce and market products that may not be halal. For example, Nimer (2002) contends that some retailers have wrongly placed halal food labels on meat items to attract Muslim customers. The Illinois act, which has the most accommodating language, specifically defines the term halal food as:

> *being prepared under and maintained in strict compliance with the laws and customs of the Islamic religion, including, but not limited to, those laws and customs of zabiha/zabeeha (slaughtered according to appropriate Islamic code) and as expressed by reliable recognized Islamic entities and scholars. [And is therefore probably unconstitutional as per the recent second circuit court of appeals ruling that was not reviewed by the U.S. Supreme Court.]*

This law makes it a misdemeanor for any person to make any oral or written statement that directly or indirectly tends to deceive or otherwise lead a reasonable individual to believe that a non-halal food or food product is halal.

In 2002, the states of California and Michigan also enacted their versions of a halal food laws. The state laws are reproduced in Appendices E (NJ), F (IL), G (MN), H (CA), and I (MI).

It is anticipated that other states with significant Muslim populations, where legislatures are sensitive to their ethnic voters, will also introduce similar bills. In early 2003, a similar bill was introduced into both the Senate and the House of the state of Texas. Unless subsequent regulations are passed to implement the laws, the passage of the bills only provides a false sense of protection and very little actual benefit is received by the Muslim consumers.

At the federal level there are no comprehensive halal laws; however, the Food Safety and Inspection Service (FSIS), U.S. Department of Agriculture, has issued a directive for the labeling of halal products (Figure 3.1). The FSIS directive states that the Food Labeling Division, Regulatory Programs, recently approved a standards and labeling policy book addition involving the use of "Halal" "Halal Style," or "Halal Brand" on meat and poultry products, provided that they are prepared under Islamic authority.

Products identified with the term halal must not contain pork or pork derivatives. The Federal Meat and Poultry Inspection does not certify the halal preparation of products, but rather accepts "halal" and similar statements, if the products are prepared under the supervision of an authorized Islamic organization. When "halal" and similar statements are used, plant management is responsible for making the identity of the Islamic organization available to inspection personnel.

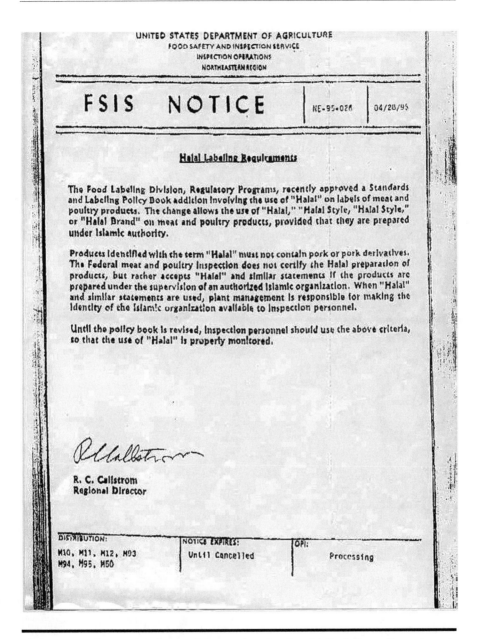

UNITED STATES DEPARTMENT OF AGRICULTURE
FOOD SAFETY AND INSPECTION SERVICE
INSPECTION OPERATIONS
NORTHEASTERN REGION

FSIS NOTICE
| NE-95-028 | 04/28/95 |

Halal Labeling Requirements

The Food Labeling Division, Regulatory Programs, recently approved a Standards and Labeling Policy Book addition involving the use of "Halal" on labels of meat and poultry products. The change allows the use of "Halal," "Halal Style, "Halal Style," or "Halal Brand" on meat and poultry products, provided that they are prepared under Islamic authority.

Products identified with the term "Halal" must not contain pork or pork derivatives. The Federal meat and poultry inspection does not certify the Halal preparation of products, but rather accepts "Halal" and similar statements if the products are prepared under the supervision of an authorized Islamic organization. When "Halal" and similar statements are used, plant management is responsible for making the identity of the Islamic organization available to inspection personnel.

Until the policy book is revised, inspection personnel should use the above criteria, so that the use of "Halal" is properly monitored.

R. C. Callstrom
Regional Director

DISTRIBUTION:	NOTICE EXPIRES:	OPI:
M10, M11, M12, M03 M94, M95, M60	Until Cancelled	Processing

Figure 3.1 The Food Safety and Inspection Service, United States Department of Agriculture, directive for the labeling of halal products. (From U. S. Department of Agriculture, Washington, D.C.)

REFERENCES

Al-Quaderi, J. 2001. Change of state — Istihala. *The Halal Consum.,* No. 2, 7.

Australian Federation of Islamic Council (AFIC). 2003. www.afic.com.au, AFIC, Waterloo, D.C., NSW, Australia.

Chaudry, M.M. 1997. Islamic foods move slowly into marketplace (press article), *Meat Process.,* 36(2), 34-38.

Food Safety and Inspection Service, United States Department of Agriculture. 1995. Inspection Operations, Northeastern Region, Halal Labeling Requirements, NE-95-06.

Khan, G.M. 1991. *Al-Dhabh: Slaying Animals for Food the Islamic Way,* Abdul-Qasim Bookstore, Jeddah, Saudi Arabia.

Nimer, M. 2002. *The North American Muslim Resource Guide,* Routledge, New York, pp. 123-124.

Riaz, M.N. 1997. Alcohol: the myths and realities, in *Handbook of Halal and Haram Products,* Uddin, Z., Ed., Publishing Center of American Muslim Research & Information, Richmond Hill, NY, pp. 16-30.

Riaz, M.N. 2000. How cheese manufacturers can benefit from producing cheese for halal market. *Cheese Mark. News,* 20(18), 4, 12.

4

INTERNATIONAL AND DOMESTIC TRADE IN HALAL PRODUCTS

HALAL MARKETS AND DEMOGRAPHICS

The estimates of the number of Muslims in 2002 vary from 1.2 billion to 1.5 billion. According to a conservative estimate, one sixth to one seventh of the world's people is Muslims (Waslien, 1992). According to the Center for American Muslim Research and Information, New York, one fourth to one fifth of the world is Muslim. A figure of 1.3 billion quoted by Chaudry (2002) seems to be the best estimate. In 1976, 57 countries were reported to have a Muslim majority, accounting for about 800 million people in 1980 (Noss, 1980), which rose to over 1 billion in 2002.

With expanding global markets, innovative food companies are leading the charge by carving a new niche to gain a competitive edge in the marketplace. For the domestic marketer the question is "How can my company get the additional edge in the marketplace?" whereas for the global marketer, the question is "How can I comply with the import requirements of the Muslim countries?" The halal food market potential in the world is not limited to Muslim countries. Countries such as Singapore, Australia, New Zealand, and South Africa (with very small Muslim populations) have become significant contributors to the world halal trade.

The world is now much more accessible because of improved transportation and communication systems. It has become truly a global supermarket. The demand for halal foods and products in countries around the world is on an increase, as Muslim consumers are creating an educated demand for halal foods and products. In the past, many Muslim countries met most of their food requirements domestically or imported them from other Muslim countries. However, population increases are outpacing food supply, and Muslim countries now import food from agriculturally advanced countries. Changes in the food habits of people, where Western style franchised food is becoming popular, is another factor in the changes taking place in international marketing.

U.S. HALAL MARKET

Muslims in North America have increased in number since World War I and even faster after World War II. As of 1992, their numbers are estimated at 6 to 8 million and they are spread all over the continent in cities both large and small (Sakr, 1993). The majority of Muslims are immigrants and their descendants. Americans and Canadians who have accepted Islam are about 25 to 30% of the current Muslim population (Sakr, 1993).

In the U.S., the top ten major halal food markets are (1) New York City, New Jersey, and Long Island metropolitan areas; (2) Los Angeles; (3) Chicago; (4) Detroit; (5) Houston; (6) Dallas/Forth Worth; (7) South Florida; (8) San Francisco; (9) Atlanta; and (10) Washington, D.C.

According to one estimate, the buying power for food of Muslim consumers in North America was worth $12 billion in 1999 (Riaz, 1999). It is estimated that amount of spending by Muslims on food will exceed $15 billion in 2003, taking into consideration the growth in Muslim population through birth and recent immigration. Sales of halal foods are gaining popularity at grocery stores and restaurants. According to one estimate, sales of halal products will continue to climb as the number of U.S. Muslims grows. Sales were up 70% in the past five years for halal meat (*The Wall Street Journal*, 1998). A Kansas company launched a line of halal frozen pizzas. An Ontario company has an extensive line of frozen halal meat, poultry, and other products. Most meat companies had been making halal food products only for export; now small- to mid-size companies are producing halal products exclusively for the domestic U.S. and Canadian markets. For the first time in the history of the U.S. military, certified halal meals are available to Muslim soldiers.

In the U.S., there are a number of areas where providers of halal products are becoming active. Among these are public and private schools, prison systems, the vending industry, and convenience foods such as frozen dinners and airline meals. The number of Muslim students in public schools is increasing annually. These children generally bring their own lunches to school or skip lunch. When vendor contracts are awarded, those who can supply halal meals to the school system will have an advantage over those who cannot supply these special meals.

As with everyone else, Muslims have a very busy life style. Increasingly, daily activities at work and home do not allow them to prepare meals at home. Muslims, like any other segment of population, are involved with their jobs and other away-from-home activities and demand on their time has increased. Availability of prepared convenience foods has become more and more important. Consequently, availability of halal prepared foods will serve a very useful purpose.

Vending is another area where a tremendous opportunity exists for halal food. Items such as sandwiches, hot meals (such as beef stew and soups), cookies, cakes and rolls, ice cream, and candy can be a very practical means of making halal foods available in places where the concentration of Muslims might not justify other types of foods. Places such as hospitals, where many Muslim doctors practice and where Muslim visitors come to check on friends, will benefit from this service. Halal food vending can also be considered for cafeterias in schools, colleges, and places of business with lower concentrations of Muslims.

There are over 8500 grocery items on the typical shelves of North American and European supermarkets, and many more are being added daily. Very few food items in grocery stores have halal markings. Muslims are making their decisions based on the ingredient information on the labels that might indicate whether that particular food item is lawful for Muslim consumption. It will be helpful for Muslim consumers to have halal markings on the label.

GLOBAL HALAL MARKET

There are ca. 1.3 billion Muslims in the world and 1.5 billion halal consumers, which means that one out of every four human beings consumes halal products. The difference of 0.2 billion between the halal consumers and Muslims is accounted for by non-Muslims living in Muslim-majority countries where most foods are halal, such as Indonesia and Bangladesh.

Presently, Southeast Asia and the Middle East are the two strong markets for halal products (Riaz, 1998). All major U.S. poultry processors export to these markets, whereas secondary suppliers provide beef. The primary sources of beef in those markets are imports from Australia and New Zealand, whose governments are very supportive of halal programs (Chaudry, 1997). Marketing efforts to supply certified halal products throughout the world are gaining momentum.

Southeast Asia is home to more than 250 million halal consumers. Indonesia, Malaysia, and Singapore have had regulations to control the import of halal-certified products for some number of years. Recently, Thailand, the Philippines, and other countries have realized the value of halal-certified products and their governments are formulating regulations to promote both export and import of halal-certified products. For export to many of the Association of South East Asian Nations (ASEAN), even the simplest of vegetable products must be certified. In this region, even non-Muslim consumers perceive halal as a symbol of quality and wholesomeness.

Middle Eastern countries are net importers of processed foods both for the food service and retail markets. Saudi Arabia, United Arab Emirates, and other Middle Eastern countries have been importing food for decades. North Africa and other African countries also offer opportunities for export of processed food as their economies and political conditions improve.

South Asia, comprising India, Pakistan, Bangladesh, and Sri Lanka, is home to almost 1.3 billion people, of which over 400 million are Muslims. Although this region is an agricultural economy, these countries do import certain processed items, especially for food service.

In the late 1980s and early 1990s, the potential of the Southeast Asian and Middle Eastern markets for halal foods started to be realized, leading to an increase in the production and certification of halal foods. This has expanded into South Asia, the Mediterranean, Europe, and Central Asia. The benefits of trade for Western corporations with Muslim-majority countries are clear. Even Muslim-minority countries such as Singapore and South Africa have shown that the halal food business is good business. Although the Muslim community forms only 16% of the 3.8 million population of Singapore, the halal food industry is big business in this cosmopolitan city. McDonald's, A & W, Kentucky Fried Chicken, and Taco Bell are some international brands that have gone 100% halal in Singapore.

The opportunities available to a corporation able to supply halal products are continuously growing. Muslims are starting to blend the best of Western attitudes with their generally Eastern cultures. Additionally, the large addition of Westerners to the faith of Islam is resulting in some changes in the behavior of the Muslim community. Whereas in the past Muslims simply avoided foods that did not meet the dietary standard of halal, today Muslims are making their presence felt socially and politically. Muslims are now requesting food products that meet their dietary needs. They are offering services and cooperating with producers with the foresight and wisdom to cater to the Muslim consumer.

Table 4.1 shows total population and Muslim population of different countries. The notes given in Table 4.1 relate to halal production and consumption in several countries. Many countries with percentage-wise lower Muslim populations have been included in this table due to their relevant importance as producers or exporters, or both, of halal foods, among them are the U.S., Canada, Argentina, Brazil, Australia, New Zealand, and France.

According to *Encyclopedia Britannica*, the countries of Benin, Cameroon, Gabon, Guinea-Bassau, Guyana, Kazakhstan, Mozambique, Togo,

Table 4.1 Halal Activity in Various Countries

Country	Population (in millions)	Muslims (%)	Number of Muslims (in millions)	Notes
Afghanistan[a]	26.88	99	26.61	1, 6
Albania[a]	3.49	70	2.44	1, 6
Algeria[a]	31.20	99	30.88	2
Argentina	37.48	2	0.75	2, 4
Australia	19.36	1	0.19	4, 5
Azerbaijan[a]	8.10	93	7.53	1
Bahrain[a]	0.70	100	0.70	3, 7
Bangladesh[a]	131.27	88	115.52	7
Benin[a]	6.59	20	1.32	1
Bosnia	2.00	60	1.20	1, 6
Brazil	172.12	0.10	0.17	2, 4
Brunei-Darussalam[a]	0.34	64	0.22	3, 7
Burkina-Faso	12.27	50	6.13	1
Cameroon[a]	15.80	21	3.32	1
Canada	31.08	1	0.31	2, 4
Chad[a]	8.42	50	4.21	1
China, P.R.	1274.91	2	25.50	2, 4
Comoros[a]	0.56	98	0.55	1
Cote d'Ivoire[a]	16.39	60	9.83	1
Djibouti[a]	0.46	97	0.45	1
Egypt[a]	68.36	90	61.52	3, 7
Fiji	0.80	20	0.16	3, 7
France	59.09	7	4.14	2, 4
Gabon[a]	1.21	3	0.04	1
Gambia[a]	1.41	95	1.34	1
Guinea[a]	7.61	85	6.47	1
Guinea-Bissau[a]	1.32	45	0.59	1
Guyana[a]	0.77	12	0.09	1
India	1029.99	15	154.50	2, 4
Indonesia[a]	224.78	87	195.56	2, 3, 5, 7
Iran[a]	65.62	99	64.96	1, 2, 3
Iraq[a]	23.33	97	22.63	1
Jordan[a]	5.13	96	4.92	1
Kazakhstan[a]	16.73	47	7.86	1
Kuwait[a]	2.27	85	1.93	2, 3, 5, 7
Kyrghyz Republic[a]	4.93	75	3.70	1
Lebanon[a]	3.63	55	2.00	1

(continued)

Table 4.1 Halal Activity in Various Countries (continued)

Country	Population (in millions)	Muslims (%)	Number of Muslims (in millions)	Notes
Libya[a]	5.24	97	5.08	1
Malaysia[a]	22.60	52	11.75	2, 3, 5
Maldives[a]	0.30	100	0.30	1
Mali[a]	11.00	90	9.90	1
Mauritania[a]	2.67	100	2.67	1
Morocco[a]	30.12	98	29.52	2, 3
Mozambique[a]	19.37	20	3.87	1
New Zealand	3.86	2	0.08	2, 4
Niger[a]	10.35	90	9.32	1
Nigeria[a]	126.64	50	63.32	1
Oman[a]	2.53	90	2.28	1, 7
Pakistan[a]	144.61	97	140.27	2
Philippines	72.00	5	3.60	2, 4
Qatar[a]	0.74	95	0.70	3, 7
Russia	144.42	10	14.44	1
Saudi Arabia[a]	22.76	97	22.08	2, 3, 5, 7
Senegal[a]	10.29	92	9.47	1
Sierra Leone[a]	5.23	60	3.14	1
Singapore	3.32	15	0.49	2, 3, 5, 7
Somalia[a]	7.49	100	7.49	1
South Africa	43.58	2	1.05	2, 4, 5
Sudan[a]	36.08	70	25.26	1
Suriname[a]	0.44	20	0.09	1
Syria[a]	16.73	90	15.06	1
Tajikistan[a]	6.44	80	5.15	1
Togo[a]	5.15	15	0.77	1
Tunisia[a]	9.83	99	9.72	1
Turkey[a]	66.23	99	65.57	1
Turkmenistan[a]	5.46	89	4.86	1
Uganda[a]	23.98	16	3.84	1
U.K.	59.95	2	1.20	2, 4
United Arab Emirates[a]	3.11	96	2.99	3, 5, 7
U.S.A.	286.07	2	5.72	2, 4
Uzbekistan[a]	25.15	88	22.13	1
Yemen[a]	18.08	100	18.06	1

(continued)

Table 4.1 Halal Activity in Various Countries (continued)

Note: Numbers in the Notes column are explained below:

1. Countries that do not have any significant international trade in food by way of export or import.
2. Countries that have strong halal activity either in food processing or export/import trade, evidenced by halal certification within that country.
3. Countries that require halal certificates for importing into that country either meat products only or meat, food, and kindred products.
4. Countries that are major exporters of food directly to Muslim countries.
5. Countries that have very organized halal certification either supported by the community or by their respective governments.
6. Countries that receive humanitarian rations or nutritional supplements under UNICEF programs.
7. Countries where processed food imports exceed processed food exports.

[a] Member of Organization of Islamic Conference (OIC).

Source: From *Encyclopedia Britannica Almanac 2003*, Encylopedia Britannica Staff, Chicago, IL.

and Uganda have a Muslim population of less than 50% and would not be considered Muslim countries; however, they are included here because:

- They are members of the Organization of Islamic Conference (OIC).
- Some of them have higher percentages of Muslims than reported by *Encyclopedia Britannica*, as reported by other sources (The Islamic Foundation, 1994).

Two major countries with relatively smaller percentages of Muslims, China and India, are included in this table because total Muslim population of these countries ranks second and twelfth among the countries with highest population (*Encyclopedia Britannica Almanac*, 2003). According to sources within China, Muslims number around 80 million in China, making it the fifth largest Muslim population (Sadek, 2003). In many cases, *Encyclopedia Britannica* reported Muslim population lower than that reported by Islamic sources, www.islamicpopulation.com.

The discussion of market activity is somewhat subjective, and individuals and companies dealing with different countries are encouraged to contact the departments of trade, health, religious affairs, etc., in the country they are interested in dealing with. One can also consult a country's import requirements, presented in Chapter 5.

Further information is compiled in Table 4.2, which covers total imports, total exports, and percentage of imported over exported food. Food manufacturing data is in the local currency of the countries for which such data is available (*Encyclopedia Britannica Almanac*, 2003).

Table 4.2 Food Imports and Exports of Selected Countries, Year 2000

Country	Imports		Exports		
	Total[a] (in millions)	Food (%)	Total[a] (in millions)	Food (%)	Food Manufacturing
Afghanistan	$525	18.8	$149	15.6 (dried foods and nuts)	
Albania	$1,070	19.8	$256	6.6	74,000 metric tons
Algeria	$9,102	27.5	$13,586	0.5 (dates)	$622 million
Argentina	$25,508	4.4	$23,333	35.1	
Australia	$A110,083	3.6	$A97,255	17.3	U.S.$12,239 million
Azerbaijan	$1,077	16.4	$606,151	7.7	AM1,972,000 million
Bahrain	BD1341	12.8	BD1,230		
Bangladesh	TK341,850		TK203,970	7.3 (fish and prawns)	
Benin	CFAF300,800		CFAF231,100		70,000 metric tons (meat)
Bosnia	$2,629		$968		
Brazil	$60,793		$51,120	20.3	R$18,117 million
Brunei	B$3,154	11.1	B$3,973		
Burkina Faso	CFAF368,700	12.3	CFAF156,600		
Cameroon	CFAF881,500	11.3	CFA993,900	15.0	CFAF49,314 million (beverages)
Canada	Can$363,281	5.4	Can$422,559	7.1	Can$52,353 million
Chad	CFAF175,000		CFAF145,300	5.8	
China	$140,166	2.7	$183,757		
Comoros	CF24,929	6.1	CF4,248		
Cote D'Ivoire	CFAF1,602,000	19.4	CFAF2,379,000		
Djibouti	$239	53.2	$59		
Egypt	$15,165	21.2	$5,327		
Fiji	F$1,393	14.8	F$715	5.3 (fish)	U.S.$84 million
France	F1,687,500	7.9	F1,773,200	12.0	

Gabon	CFAF578,100	23.1	CFAF1,776,300		
Gambia	D2,187	32.9	D81		D629 million
Guinea	$572	17.1	$709	8.7 (fish and coffee)	
Guinea-Bissau	CFAF51,800	35.1	CFAF28,300	94.4 (cashews)	CFAF10,500 million
Guyana	$55		$525	38 (sugar and rice)	
India	Rs2,006,570		Rs1,607,430		Rs139,000 million
Indonesia	$27,337	9.6	$48,848		Rp9,028,000 million
Iran	$14,286		$13,118		$1,170 million
Iraq	$2,500	42.7	$419		
Jordan	JD2,720	19.6	JD1,276		
Kazakhstan	$6,575	3.7	$5,774		107,397 Teng
Kuwait	KD2,318	14.9	KD3,696		
Kyrghyz	$709	11.7	$604	13.2	729 million som
Lebanon	LP6,228	8.8	LP714	15.1	
Libya	$5,593	20.0	$9,029		
Malaysia	RM228,309	4.6	RM286,756		
Maldives	Rf3,551	21.4	Rf699	59.0	
Mali	CFAF490,600	13.9	CFAF348,600		
Mauritania	$403	(vegetable oils and fats)	$405		
Morocco	DH95,577	17.1	DH72,283	31.6	
Mozambique	$783	18.3	$226	66.0	Mt696,611 million
New Zealand	$NZ24,248	74.0	$NZ22,600		
Niger	CFAF196,700		CFAF175,600		
Nigeria	N111,728	8.4	N220,409		N25,415 million
Oman	RO1,938	12.2	RO4,352	1.9	RO72,930 million
Pakistan	$10,361	8.9	$8,569		
Philippines	$30,723		$35,037		P246,300 million
Qatar	QAR9,098	13.2	QAR26,258		

(continued)

Table 4.2 Food Imports and Exports of Selected Countries, Year 2000 (continued)

Country	Imports Total[a] (in millions)	Imports Food (%)	Exports Total[a] (in millions)	Exports Food (%)	Food Manufacturing
Russia	$40,200		$75,100		
Saudi Arabia	SR104,980	7.3 (vegetables)	SR190,100		
Senegal	CFAF1,155,300	22.5	CFAF339,250		
Sierra Leone	Le153,856	51.6	Le11,347		
Singapore	S$232,175		S$237,826		
Somalia	$327		$187	51.4	SoSh795 million
South Africa	R98,614	19.5	R101,397		$3,028 million
Sudan	$1,412	9.7	$780		
Suriname	Sf258,917	13.8	Sf211,021		Sf992 million
Syria	LS43,010		LS38,880	1.3	
Tajikistan	$731		$586		
Togo	CFAF263,400		CFAF244,800		CFAF41,400
Trinidad	TTD$18,887	7.3%	TT$14,221		
Tunisia	D11,728		D8,005		
Turkey	$53,983		$27,324		$3,944 million
Turkmenistan	$1,137	8.0	$614		
Uganda	$1,411		$458		

U.K.	£194,434	4.7	£165,667	P19,337 million
U.A.E.	Dh109,100	9.7	Dh139,500	Dh2,122 million
U.S.	$1,025,032	3.4	$695,009	$102,103 million (meat products only)
		5.3		
Uzbekistan	$2,717	20.9	$2,888	
Yemen	$1,536	36.2	$2,436	

Note: $ is U.S. Dollar (1.00); $A is Australian Dollar (1.51 U.S.); AM is Azerbaijan Manat (4,857.00 U.S.); BD is Bahraini Dinar (0.37 U.S.); TK is Bangladeshi Taka (61.00 U.S.); £ is British Pound, U.K. (0.62 U.S.); R$ is Brazilian Real (3.07 U.S.); B$ is Brunei Dollar (1.76 U.S.); Can$ is Canadian Dollar (1.38 U.S.); CFAF is CFA Francs (664.20 U.S.); CF is Comoros Francs (432.00 U.S.); F$ is Fiji Dollar (1.88 U.S.); F is French Franc (0.88 U.S.); D is Gambian Dalasi (28.00 U.S.); Rs is Indian Rupee (45.95 U.S.); Rp is Indonesian Rupiah (8,547.00 U.S.); JD is Jordanian Dinar (0.71 U.S.); Teng is Kazakhstan Tenge (145.80 U.S.); KD is Kuwaiti Dinar (0.30 U.S.); LP is Lebanese Pound (1,549.00 U.S.); RM is Malaysian Ringit (3.79 U.S.); Rf is Maldives Rufiyaa (12.69 U.S.); DH is Moroccan Dirham(9.79 U.S.); Mt is Mozambique Metical (23,505.00 U.S.); $NZ is New Zealand Dollar (1.69 U.S.); N is Nigerian Naira (131.00 U.S.); RO is Omani Riyal (0.38 U.S.); P is Philippine Peso (55.15 U.S.); QAR is Qatari Riyal (3.64 U.S.); SR is Saudi Riyal (3.75 U.S.); Le is Sierra Leone Leone (1,930.00 U.S.); S$ is Singapore Dollar (1.75 U.S.); SoSh is Somali Shilling (2,620.00 U.S.); R is South African Rand (7.36 U.S.); Sf is Suriname Guilder (2,502.00 U.S.); LS is Syrian Pound (47.71 U.S.); TTD is Trinidad/Tobago Dollar (6.11 U.S.); Dh is UAE Dirham (3.67 U.S.).

Source: From *Encyclopedia Britannica Almanac 2003*, Encylopedia Britannica Staff, Chicago, IL.

It is interesting to note how food value as a percentage of import or export varies from country to country. For instance, 94.4% of Guinea-Bissau exports is accounted by food, with a major product being cashews. Fifty nine percent of the Maldives exports is food, mainly fish. Countries with higher imports than exports of food, at least 50% more, include Afghanistan, Albania, Azerbaijan, Bangladesh, Bosnia, Burkina-Faso, Comoros, Djibouti, Egypt, Fiji, Gambia, Iraq, Jordan, Lebanon, Maldives, Mozambique, Senegal, Sierra Leone, Somalia, Turkey, Turkmenistan, Uganda, and the U.S.

This data might not be very conclusive and completely accurate, but it offers some insight into the level of import, export, and food manufacturing in select countries. Table 4.3 lists the top 20 countries with the highest Muslim populations, representing about 90% of the Muslim population. Any country dealing with this segment has to consider meeting halal guidelines to make its food supply acceptable to Muslim consumers.

Table 4.3   Countries with Highest Muslim Population

Country	Population (in millions)	Muslims (%)	Number of Muslims (in millions)
Indonesia	224.78	87	195.56
India	1029.99	15	154.50
Pakistan	144.61	97	140.27
Bangladesh	131.27	88	115.52
Turkey	66.23	99	65.57
Iran	65.62	99	64.96
Nigeria	126.64	50	63.32
Egypt	68.36	90	61.52
Algeria	31.20	99	30.88
Morocco	30.12	98	29.52
Afghanistan	26.88	99	26.61
China	1274.91	2	25.50
Sudan	36.08	70	25.26
Iraq	23.33	97	22.63
Uzbekistan	25.15	88	22.13
Yemen	18.08	100	18.06
Syria	16.73	90	15.06
Malaysia	22.60	52	11.75
Mali	11.00	90	9.90
Cote d'Ivoire	16.39	60	9.83

Among the top four countries, Indonesia is unquestionably the leader in the amount of food imported and in its strictness in meeting halal regulations. Malaysia was the pioneer in establishing halal laws in the early 1980s and remains a force in matters relating to halal certification globally. Many of the countries produce and consume a majority of their daily food needs or may import bulk agricultural commodities for further processing.

During the past two decades, South East Asian and Middle Eastern countries have witnessed the westernization of the food service industry. Global food service giants such as McDonald's® and Kentucky Fried Chicken® are a regular feature of the landscape. The Western food companies have to comply with the local halal requirements to compete and be successful.

Table 4.4 shows U.S. exports of agricultural and fish products to the Middle East and North Africa, broken down by commodity types, for example, bulk agricultural commodities, intermediate agricultural commodities, consumer-oriented foods, and fish and seafood products from 2000 to 2002. Among the major food commodities, wheat and rice imports decreased whereas imports of coarse grains and soybean increased. The industry saw over a 50% increase in soybean trade over these three years.

Among the intermediate agricultural commodities, the import of wheat flour and soybean meal decreased whereas there was a significant increase in the import of vegetable oil, soybean oil, livestock, and sugar. Total value of commodities in this category increased from U.S.$832 million to U.S.$1.07 billion.

Major categories of food exports to the Middle East among consumer products were tree nuts, processed fruits and vegetables, dairy products, snack foods, poultry meat, red meat, fresh fruit, and breakfast cereals. Prepared meats, although a smaller-volume item, registered an increase of 400% over the two year period. Except for breakfast cereals, all other commodities gained in import values to a net increase of about 10%. Total agricultural products, excluding fish exported to this region, were valued at $4.0 billion in 2002 over $3.7 billion in 2000.

A similar trend was observed for U.S. exports of agricultural products to the ASEAN region, as evident from Table 4.5, where total value of trade of agricultural products excluding fish went up from $2.58 billion in 2000 to $2.88 billion in 2002. Looking at individual categories among all classes of commodities, the changes were more subtle, except for a decrease in rice from $71.6 million to $17.7 million and an increase in soybean from $411.3 million to $467.3 million. The import of prepared meat actually decreased from $7.2 million to $4.2 million, perhaps with a shift to local production. There also was a decrease in the amount of poultry meat imported from the U.S. during this same period.

Table 4.4   U.S. Exports (in Thousands of Dollars) of Agricultural Products to Middle East and North Africa for Fiscal Years 2000–2002[a]

Product	2000	2001	2002
Bulk Agricultural			
Total	2,555,088	2,217,676	2,397,848
Wheat	929,828	702,847	655,692
Coarse grains	963,978	904,402	1,031,567
Rice	141,946	119,566	70,993
Soybean	198,714	204,611	309,466
Cotton	184,842	185,732	275,959
Tobacco	105,780	60,908	20,049
Pulses	12,487	19,249	13,357
Peanuts	1,376	844	790
Other bulk commodities	16,138	19,518	19,976
Intermediate Agricultural			
Total	831,862	876,316	1,075,440
Wheat flour	20,240	7,772	13,130
Soybean meal	263,554	259,344	221,282
Soybean oil	19,219	31,088	78,860
Vegetable oils (excluding soybean oils)	174,520	151,495	228,594
Feeds and fodder (excluding pet foods)	49,877	63,811	54,979
Live animals	50,535	91,453	132,408
Hides and skin	10,491	12,780	22,641
Animal fats	54,592	33,444	47,176
Planting seeds	52,003	42,646	50,371
Sugars, sweeteners, and beverage bases	74,189	105,983	111,175
Other intermediate products	62,641	76,498	114,825
Consumer-Oriented Agricultural			
Total	479,541	539,171	523,157
Snack foods (excluding nuts)	44,472	58,822	60,006
Breakfast cereal and pancake mix	17,000	20,455	11,783

(continued)

Table 4.4 U.S. Exports (in Thousands of Dollars) of Agricultural Products to Middle East and North Africa for Fiscal Years 2000–2002[a] (continued)

Product	2000	2001	2002
Consumer-Oriented Agricultural (continued)			
Red meats, fresh, chilled, or frozen	40,973	42,209	50,756
Red meats, prepared or preserved	1,921	6,686	8,309
Poultry meats	40,298	45,986	46,209
Dairy products	53,993	59,889	33,234
Eggs and products	2,164	3,847	3,407
Fresh fruit	29,554	31,048	25,825
Fresh vegetables	4,444	5,323	6,914
Processed fruits and vegetables	59,951	55,330	59,967
Fruit and vegetable juices	12,533	9,787	14,464
Tree nuts	67,710	87,299	93,982
Wine and beer	6,311	6,460	6,470
Nursery products and cut flowers	1,410	505	1,124
Pet foods (dog and cat food)	6,536	10,982	10,407
Other consumer-oriented products	90,172	94,544	94,302
Fish and Seafood Products, Edible			
Total	12,309	12,364	13,715
Salmon, whole or eviscerated	318	878	671
Salmon, canned	1,735	2,277	831
Crab and crabmeat	298	173	461
Surimi (fish paste)	52	7	74
Roe and urchin (fish eggs)	1,209	1,318	1,928
Other edible fish and seafood	8,698	7,710	9,750
Agricultural product total (excluding fish)	3,866,491	3,633,163	3,996,446

[a] Fiscal years are from October to September, e.g., fiscal year 2000 is from October 1999 to September 2000.

Source: From data compiled from U.S. Department of Commerce, Foreign Trade Statistics (USDA 2003). Analysis by Commodity & Marketing Programs, Foreign Agricultural Service, USDA.

Table 4.5 U.S. Exports (in Thousands of Dollars) of Agricultural Products to ASEAN Region for Fiscal Years 2000–2002[a]

Product	2000	2001	2002
Bulk Agricultural			
Total	1,293,005	1,368,165	1,415,017
Wheat	348,671	398,976	357,518
Coarse grains	72,802	62,641	30,038
Rice	71,606	44,694	17,669
Soybean	411,275	412,498	467,317
Cotton	269,566	313,759	403,818
Tobacco	107,867	118,086	118,641
Pulses	4,802	5,888	7,301
Peanuts	126	79	142
Other bulk commodities	6,291	11,543	12,573
Intermediate Agricultural			
Total	670,847	830,406	807,877
Wheat flour	4,945	1,003	483
Soybean meal	277,520	367,681	290,394
Soybean oil	551	365	4,692
Vegetable oils (excluding soybean oils)	18,309	20,983	23,050
Feeds and fodder (excluding pet foods)	108,988	109,178	124,242
Live animals	11,125	14,230	14,811
Hides and skin	29,864	58,414	70,447
Animal fats	438	595	548
Planting seeds	8,898	4,831	4,348
Sugars, sweeteners, and beverage bases	52,122	56,104	74,178
Other intermediate products	158,088	197,022	200,683
Consumer-Oriented Agricultural			
Total	616,561	708,872	656,768
Snack foods (excluding nuts)	80,238	82,313	76,891
Breakfast cereal and pancake mix	5,624	6,384	4,875

(continued)

Table 4.5   U.S. Exports (in Thousands of Dollars) of Agricultural Products to ASEAN Region for Fiscal Years 2000–2002[a] (continued)

Product	2000	2001	2002
Consumer-Oriented Agricultural (continued)			
Red meats, fresh, chilled, frozen	25,691	25,436	25,357
Red meats, prepared or preserved	7,242	5,418	4,203
Poultry meats	52,678	38,277	38,333
Dairy products	83,857	127,432	84,757
Eggs and products	2,123	2,132	2,266
Fresh fruit	103,822	160,743	156,473
Fresh vegetables	12,171	12,222	8,604
Processed fruits and vegetables	104,806	108,300	105,398
Fruit and vegetable juices	11,089	13,539	18,648
Tree nuts	12,497	15,440	15,179
Wine and beer	10,051	9,560	9,759
Nursery products and cut flowers	674	137	167
Pet foods (dog and cat food)	16,531	19,868	18,818
Other consumer-oriented products	87,466	81,671	87,041
Fish and Seafood Products, Edible			
Total	40,014	57,967	58,189
Salmon, whole or eviscerated	7,343	14,175	16,010
Salmon, canned	1,289	556	475
Crab and crabmeat	1,179	1,257	2,044
Surimi (fish paste)	683	618	977
Roe and urchin (fish eggs)	675	1,689	7,997
Other edible fish and seafood	28,845	39,672	30,686
Agricultural product total (excluding fish)	2,580,413	2,907,443	2,879,663

[a] Fiscal years are from October to September, e.g., fiscal year 2000 is from October 1999 to September 2000.

Source: From data compiled from U.S. Department of Commerce, Foreign Trade Statistics (USDA 2003). Analysis by Commodity & Marketing Programs, Foreign Agricultural Service, USDA.

Table 4.6 shows similar data on exports to South Asia, which includes Pakistan, India, Bangladesh, Sri Lanka, and other smaller countries. This region has a very large total and Muslim population but imports fewer commodities compared to the Middle East and South East Asia. Still, an increase of 80% from 2000 to 2002 is shown, but much of this increase is in nonfood areas, especially cotton. Wheat, oil, and nuts are imported in the largest quantities into this region. Import of wheat actually declined from $118 million in 2000 to $95 million in 2002, whereas export of vegetable oil and soybean oil combined increased from $26 million to $105 million over the same period.

Comparison of Table 4.4, Table 4.5, and Table 4.6 shows that the Middle East imported more agricultural commodities in value than the ASEAN region, with South Asia being far behind. The ASEAN region, however, imported more fish and seafood with $58 million, compared to $13 million for the Middle East and less than $1 million for South Asia. The Middle East has been a strong importer of poultry products from many Western countries, with Brazil and France being the leading exporters.

From the export data it seems that the South Asian region is a sleeper. As the taste of people changes from local cuisine to international menus of the burgers, fried chicken, and French fries, this region will see an explosion of imports and establishments of global restaurant chains. The other significant areas where activity in the import of processed foods is expected to grow are China and the central Asian countries.

Table 4.6   U.S. Exports (in Thousands of Dollars) of Agricultural Products to South Asia for Fiscal Years 2000–2002[a]

Product	2000	2001	2002
Bulk Agricultural			
Total	226,072	355,780	450,890
Wheat	118,021	79,712	95,297
Coarse grains	1,818	33	94
Rice	911	558	1,218
Soybean	0	7,945	24,457
Cotton	93,747	249,669	309,595
Tobacco	5,221	10,227	5,139
Pulses	6,033	7,212	12,852
Peanuts	0	4	0
Other bulk commodities	320	370	2,238

(continued)

Table 4.6   U.S. Exports (in Thousands of Dollars) of Agricultural Products to South Asia for Fiscal Years 2000–2002ᵃ (continued)

Product	2000	2001	2002
Intermediate Agricultural			
Total	113,185	131,513	207,010
Wheat flour	10,390	6,142	7,095
Soybean meal	0	0	0
Soybean oil	16,876	42,596	94,621
Vegetable oils (excluding soybean oils)	9,211	2,827	20,775
Feeds and fodder (excluding pet foods)	1,270	3,415	3,118
Live animals	1,633	1,558	1,124
Hides and skin	1,106	3,438	2,057
Animal fats	6,209	26	4,112
Planting seeds	8,996	10,786	8,007
Sugars, sweeteners, and beverage bases	1,819	2,996	902
Other intermediate products	55,677	57,729	65,199
Consumer-Oriented Agricultural			
Total	75,718	82,821	92,551
Snack foods (excluding nuts)	2,002	1,768	1,585
Breakfast cereal and pancake mix	114	534	411
Red meats, fresh, chilled, or frozen	109	131	173
Red meats, prepared or preserved	3	26	37
Poultry meats	819	581	681
Dairy products	2,169	1,050	2,515
Eggs and products	569	308	328
Fresh fruit	1,896	6,373	10,784
Fresh vegetables	257	34	87
Processed fruits and vegetables	2,825	2,792	1,955
Fruit and vegetable juices	874	895	618

(continued)

Table 4.6 U.S. Exports (in Thousands of Dollars) of Agricultural Products to South Asia for Fiscal Years 2000–2002ᵃ (continued)

Product	2000	2001	2002
Consumer-Oriented Agricultural (continued)			
Tree nuts	58,536	62,045	67,006
Wine and beer	330	237	252
Nursery products and cut flowers	222	161	77
Pet foods (dog and cat food)	105	141	355
Other consumer-oriented products	4,888	5,746	5,685
Fish and Seafood Products, Edible			
Total	949	105	626
Salmon, whole or eviscerated	0	0	99
Roe and urchin (fish eggs)	0	0	0
Other edible fish and seafood	949	105	527
Agricultural product total (excluding fish)	414,975	570,115	750,451

ᵃ Fiscal years are from October to September, e.g., fiscal year 2000 is from October 1999 to September 2000.

Source: From data compiled from U.S. Department of Commerce, Foreign Trade Statistics (USDA 2003). Analysis by Commodity & Marketing Programs, Foreign Agricultural Service, USDA.

REFERENCES

Anon. A special background report on trends in industry and finance, *The Wall Street Journal*, March 5, 1998, Sec. A, p. 1.

Chaudry, M.M. 1997. Islamic foods move slowly into marketplace (press article), *Meat Proc.*, 36(2), 34-38.

Chaudry, M.M. 2002. Halal certification process, presented at *Market Outlook: 2002 Conference, Toward Efficient Egyptian Processed Food Export Industry in a Global Environment*, Cairo, Egypt.

Encyclopedia Britannica Almanac 2003, Encylopedia Britannica Staff, Chicago, IL.

Noss, J.B. 1980. *Man's Religion*, 6th ed., Macmillan, New York, p. 15.

Riaz, M.N. 1998. Halal food: an insight into a growing food industry segment, *Food Market. Technol.*, 12(6), 6-9.

Riaz, M.N. 1999. Examining the halal market, *Prep. Foods*, 168(10), 81-85.

Sadek, M. 2003. Personal communication.

Sakr, A.H. 1993. Current issues of halal foods in North America, *Light*, May/June, 3(3), 22-24.

The Islamic Foundation. 1994. *The Muslim World Map*, Leicester, U.K.

USDA. 2003. U.S. Exports of Agricultural Products. www.fas.USDA.gov/ustrade/USTEx-BICO.asp?QI=.

Waslien, C.I. 1992. Muslim dietary laws, nutrition, and food processing, in *Encyclopedia of Food Science and Technology*, Vol. 2, Hui, Y.H., Ed., John Wiley & Sons, New York, pp. 1848-1850.

5

IMPORT REQUIREMENTS FOR DIFFERENT COUNTRIES

INTRODUCTION

International halal food trade, worth approximately $150 billion (Egan, 2002), covers many countries. Several Muslim countries are net importers of foods processed primarily in North America, Europe, Australia, and New Zealand. With the advancement in technology, particularly in food production, brand identity, franchising, and transportation, more processed foods are entering international trade than ever before. Global food giants as well as mid- to small-size companies are involved in producing and marketing their products to Muslim consumers worldwide, not only in Muslim-majority countries but also to consumers elsewhere. However, most of the companies lack a clear understanding of halal requirements and import regulations of Muslim countries. They usually struggle in trying to meet the mandated requirements.

Hence, the food industry needs to understand the requirements for producing products for Muslim markets. It also needs to understand the import requirements of countries with Muslim populations, which cover religious as well as safety aspects of imported food. In this chapter we explain the import requirements of many countries that import food products, especially meat and poultry, from the U.S.

Many countries have passed laws and established halal guidelines not only for imported products but also for food products manufactured and offered for sale domestically. The requirements are essentially the same for imported and domestic products, but the method of implementation varies. Halal activity in leading Islamic countries and regions is presented.

MALAYSIA

Malaysian Muslim consumers became exposed to imported food products in the 1970s when global food service establishments started opening restaurants there. Consumers wanted an assurance that the food offered at restaurants as well as in stores was indeed halal. This prompted the Malaysian government to enact laws as well as devise procedures and guidelines with regards to halal foods, domestic and imported. The passage of the Trade Description (use of expression "halal") Order of 1975 made it an offense to falsely label food as halal, and the Trade Description Act (halal sign marking) of 1975 made it an offense to falsely claim the food to be halal on signs and other markings. A two-tier punishment was established under both acts, one for small unincorporated businesses and the other for corporations. For small businesses, the penalties established for any such offense under both of these acts is a fine not to exceed RM 100,000 (U.S.$26,000) or jail for not more than 3 years for the first offense and a fine not to exceed RM 200,000 (U.S.$52,000) or jail for not more than 6 years for the second offense. For corporations, the fines are not to exceed RM 250,000 (U.S.$65,000) for the first offense and a fine of RM 500,000 (U.S.$130,000) for each additional offense.

Over the next several years, compliance and halal certification started to take shape. The Malaysian government established a committee on evaluation of foods, drinks, and goods utilized by Muslims under the Islamic Affairs Division in the Prime Minister's Department in 1982. This committee has been solely responsible for checking and instilling halal awareness amongst food producers, distributors, and importers. Also in 1982, the Malaysian government issued regulations making it mandatory for all meat (beef, mutton, veal, and poultry) imported into Malaysia to be halal certified and such meat to originate only from meat plants approved by the Islamic Affairs Division of the Prime Minister's Department and the Department of Veterinary Services, Malaysia. The Islamic Affairs Division was later elevated to the status of a department, called the Islamic Affairs Department, separate from the Prime Minister's Department. This new entity responsible for monitoring halal matters is called Jabatan Kemajuan Islam Malaysia (JAKIM).

Under the Malaysian regulations, all halal certificates for meat and poultry must be issued and signed by an Islamic center accredited to do halal certification by JAKIM. Moreover, slaughterhouses for producing such meat and poultry products must also be approved by both Malaysian government agencies (JAKIM and the Department of Veterinary Services). Malaysian general guidelines on the slaughtering of animals and the preparation and handling of halal food are given in Appendix C.

For processed food products, halal certificates issued by recognized Islamic organizations in exporting countries are sufficient. However, if a company wishes to use the official Malaysian halal logo, the processing facility in the country of origin has to be inspected and evaluated for halal certification by a team of two auditors from JAKIM.

SINGAPORE

The Islamic Council of Singapore [Majlis Ugama Islam Singapura (MUIS)] started to provide halal services in 1972 and the first halal certificate was issued in 1978. In Singapore, all imported meat (including poultry) and meat products must be halal certified by an Islamic organization in the country of export and approved by MUIS.

MUIS is solely responsible and performs a regulatory function in halal under authorization from the government of Singapore. It facilitates halal food trade through the following activities:

- Certifying local exporters to export their products to a global halal market
- Certifying local establishments
- Participating in forums on standardization of halal certification

The Singapore parliament passed the amendment of the Administration of Muslim Act (AMLA) in 1999. This amendment gave MUIS more powers by allowing it to regulate, promote, and enhance the halal business.

In Singapore, three government agencies work with MUIS in halal enforcement: The Food Control Department, Ministry of Environment; the Agro-Veterinary Authority, Ministry of National Development; and the Commercial Crime Department, Ministry of Home Affairs.

Singapore is also a member of the ASEAN Adhoc Working Group, committee on halal food guidelines. Singapore's halal regulations are given in Appendix D.

INDONESIA

A program of halal verification in Indonesia started under the auspices of the Religious Council of Indonesia, locally known as Majelis Ulama Indonesia (MUI), has assigned the responsibility to coordinate halal activities to an institute called the Assessment Institute for Foods, Drugs and Cosmetics (AIFDC). The institute is known in the local language as Lembaga Pengkajian Pangan, Obat-obatan dan Kosmetika (LP-POM).

Halal certification within Indonesia is a multidisciplinary function. The process of inspection and evaluation is initiated with an application by the manufacturer or the importer. A team comprising auditors from LP-POM, the Department of Health, and the Department of Religious Affairs makes an inspection visit to the manufacturing facility to evaluate the production procedures and ingredients. The findings of the visit are submitted to a committee of auditors who prepare a report for the Fatwa Commission of MUI, which declares the product to be halal or not. The Department of Health is also responsible for approval of labels.

For imported food products, a team of three LP-POM auditors visits to evaluate the production facility in the country of origin and products produced therein. If the food products meet the halal requirements, a report is submitted as given previously and a halal certificate is issued. Indonesia maintains a list of approved halal-certifying agencies in many countries, including the U.S. The list of approved certifiers in the U.S. appears in Chapter 19.

MIDDLE EAST COUNTRIES

Gulf Standards

Gulf countries in the Middle East under the leadership of Saudi Arabia have formulated quite elaborate standards for meat and prepared foods to be used by the member countries. The standard is known as the Gulf Standard and Saudi Standard. It includes guidelines and requirements for the import of a variety of food and food products and meat and meat products.

All countries in the Middle East, Gulf, and the rest require that imported products be accompanied with a halal certificate issued by a recognized Islamic organization in the country of export. They also require that the certificates be endorsed by the National U.S. Arab Chamber of Commerce or the consulate of the importing country before exports can commence.

SOUTH ASIA

The South Asian countries of India, Pakistan, Bangladesh, and Sri Lanka also import halal products for consumption, especially for food service. Halal programs in these countries are not as well defined. Some of the food service establishments in Pakistan operate under guidelines similar to those in Malaysia and voluntarily require halal certificates from their vendors not only for meat and poultry but also for processed food items.

OTHER COUNTRIES

Many other countries may also require halal certificates to accompany imported food products either under a formal program or informally.

These countries include Egypt, Iran, Turkey, Thailand, Philippines, South Africa, and Australia.

REVIEW OF THE HALAL FOOD INDUSTRY

Several governments and nongovernmental organizations (NGOs) are involved in halal evaluation, halal monitoring services, halal enforcement, disseminating information, issuing halal certificates, and authorizing the use of halal logos and markings to ensure that foods are imported or produced domestically.

Many countries have published their requirements for the import of products. These requirements are made available to the countries of export. The import requirements are available from the United States Department of Agriculture (USDA). Import requirements for Bahrain, Egypt, Indonesia, Iraq, Malaysia, Oman, Pakistan, Qatar, Saudi Arabia, Singapore, Turkey, and United Arab Emirates are given in Appendix J.

REFERENCE

Egan, M. 2002. An overview of halal from the Agri-Canada perspective, paper presented at the *4th International Halal Food Conference: Current and Future Issues in Halal,* Toronto, Canada, April 21-23.

6

HALAL PRODUCTION REQUIREMENTS FOR MEAT AND POULTRY

Among all the dietary restrictions or prohibitions placed on Muslims by God, the majority fall in the domain of animal kingdom, especially land animals (see Chapter 2). In addition, Muhammad emphasized certain conditions for the handling of animals. He said, "Verily Allah has prescribed proficiency in all things. Thus, if you kill, kill well; and if you perform dhabiha, perform it well. Let each one of you sharpen his blade and let him spare suffering to the animal he slays" (Khan, 1991). In this chapter on halal production requirements, guidelines and procedures have been formulated keeping dietary permissions and prohibitions in mind.

For commercially processed poultry, birds are generally acquired from poultry farms that raise chickens specifically for that purpose, or hens and roosters may be acquired from poultry farms that raise chickens for eggs when their egg production decreases below a certain level. Chickens of any size, age, and gender may be used for halal production depending on the end use. Hens and roosters are used for high-temperature cooking, such as canning, retorting, or even dehydrating for incorporation into soups and other dry blends. In the Middle East, smaller and younger birds (ca. 3 lb) are preferred because they are used for roasting on rotisseries. The preferred feed for halal poultry should be devoid of any animal by-products or other scrap materials, which is a common practice in the West. Some halal slaughterhouses use an integrated approach, for example, where they raise their own chickens on clean feed, but most halal processors do not exert any influence over the feed. Muslim retailers then prefer free-range farmed chickens, such as those produced by Amish people, who do not feed animal by-products to the birds. However, these birds are quite large and may be best used for whole cut-up chicken or for individual parts. The use of hormones in chickens for egg or meat production is discouraged; some scholars call it haram whereas others ignore it.

METHODS OF SLAUGHTERING FOR MEAT AND POULTRY

The traditional method of slaughtering in Islam is to slit the throat, cutting the carotid arteries, jugular veins, trachea, and the esophagus, without severing the head. It must be done by a Muslim of sound mind and health while pronouncing the name of God on each animal or bird. To carry out the slaughtering process properly by hand, a team of Muslim slaughter persons is required at each line. The number of slaughter persons depends on line speed, size of the animals, and number of hours the operation will be performed. Slaughtering by hand is still preferred by all Muslims and quite widely followed in Muslim countries and other countries where Muslims control slaughterhouses.

Mechanical or machine slaughter of birds, which was initiated in Western countries, is gaining acceptance among Muslims. Almost all countries that import chicken accept machine-killed birds. The method of slaughter by machine approved by the Malaysian government is different in the following aspects from what is usually practiced in the industry in the West:

■ A Muslim while pronouncing the name of God switches on the machine.
■ One Muslim slaughter person is positioned after the machine to make a cut on the neck if the machine misses a bird or if the cut is not adequate for proper bleeding. In commercial poultry processing, generally the machine does not properly cut 5 to 10% of the birds. A Muslim then cuts the missed birds. The Muslim back-up slaughter person also continuously invokes the name of God on the birds while slaughtering and witnessing the machine kill. The height of the blade is adjusted to make a cut on the neck, right below the head, and not across the head. A rotary knife should be able to cut at least three of the passages in the neck. Any birds that are not properly cut may be tagged by the Muslim slaughter person/inspector, and used as non-halal. Two slaughter persons might be required to accomplish these requirements, depending on the line speed and efficiency of the operation.
■ The machine must be stopped during breaks and restarted by the previous procedure.

Whether slaughtered by hand or machine, birds must be completely lifeless before they enter the defeathering area. The conditions for defeathering, such as water temperature and chlorine level, are the same for halal processing as for regular poultry processing. However, in poultry processing plants where both halal and non-halal birds are processed, halal birds must be completely segregated during defeathering, chilling, eviscerating, processing, and storing. It is a common practice to chill the birds in cold water, where they might pick up water in varying percentages. Air chilling rather than chilling with water used by some companies is preferred over water chilling. Containers with halal products should be stamped halal with proper codes and markings by the authorized halal inspector. A halal certificate issued by the halal inspector in charge of the facility must accompany halal-processed items when they are shipped to another facility for further processing. Further processing, such as marinating, breading, and application of batters or rubs, should also be done under the supervision of a qualified halal inspector by using thoroughly cleaned equipment. Nonmeat ingredients such as spices, seasonings, and breadings must also be halal approved (Regenstein and Chaudry, 2001).

HALAL CONTROL POINTS FOR SLAUGHTERHOUSES

Halal control points can be determined for each operation from raising the animals to the final packaged product offered for sale. Figure 6.1 shows halal control points (HCPs) in meat and poultry processing.

HCP1

The animal must be acceptable halal species such as sheep, lambs, goats, cows, bulls, steers, heifers, broiler chickens, hens, roosters, ducks, turkeys, quails, or pigeons. Pigs, boars, swine, dogs, cats, lions, cheetahs, bears, falcons, eagles, vultures, and the like cannot be considered halal even if they are slaughtered in a halal manner.

HCP2

Islam advocates merciful treatment of animals. Hence, animals must be treated such that they are not stressed or excited prior to slaughter. Holding areas for cattle should be provided with drinking water. Excessive use of electric prods or sticks must be avoided if the animals get overly excited with such tools. Animals should be nourished and well rested. For proper preslaughter handling of animals as well as restraining for slaying, use of ritual slaughter guidelines is recommended (Regenstein and Grandin, 2002). Detailed ritual slaughter guidelines are given in Appendix K.

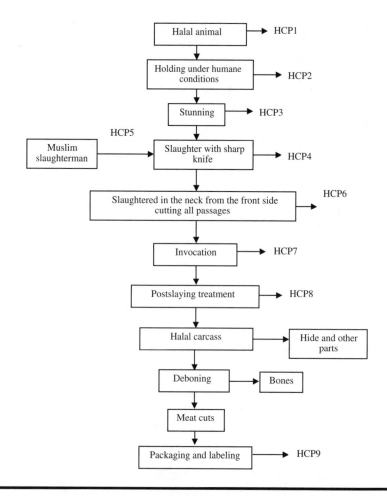

Figure 6.1   Halal control points (HCPs) in meat and poultry processing.

HCP3: Stunning

It is preferable that animals be slaughtered without stunning but with proper humane restraining systems. However, nonlethal methods of stunning might be used to meet the legal requirement for humane slaughter regulations. The animal must be alive at the time of slaying and must die of bleeding rather than a blow or electrocutions.

Captive Bolt Stunning

A mechanical stunner where a bolt enters the head and then retracts, making the animal unconscious, is acceptable. The animal while unconscious is hung by one leg and slayed within 1 to 3 min, depending on the size of the animal. The blow to the head and the timing of slaying must be adjusted so that the animal does not die before slaying.

Electric Stunning

Low-voltage stunning for a short time is not lethal to animals. Whether electric stunning is painful to the animal has not been adequately studied and will not be discussed in this section.

Different conditions are used for electrical stunning, depending on the region of the world. Although poultry is not required by law to be stunned before slaughter in the U.S., virtually all commercial poultry is stunned for humane, efficiency, and quality reasons. The birds receive 10–20 mA per broiler and 20–40 mA per turkey for 10 to 12 sec. These conditions yield adequate time of unconsciousness for the neck to be cut and sufficient blood to be lost so as to kill the bird before it regains consciousness. In most European countries, laws require poultry to be stunned, and with much higher amperages (90+ mA per broiler and 100+ mA per turkeys for 4 to 6 sec). These laws and high amperages are intended for humane treatment to ensure that the birds are irreversibly stunned so that there is no chance they will be able to recover and sense any discomfort. Essentially, these European electrical stunning conditions kill the bird by electrocution and cardiac arrest, stopping blood flow to the brain. Thus, death is by loss of blood supply to the brain for both stunning conditions, but one is by removal of blood and the other is by stopping blood flow to the brain. The harsher European electrical conditions also result in higher incidences of hemorrhaging and broken bones (Sams, 2001).

Use of low voltage is recommended for halal slaughtering, because low-voltage stunning does not kill the birds. Voltage higher than 40 mA for turkeys and higher than 20 mA for broilers must be avoided. Each plant must establish its own working procedures depending on the size and age of the birds, so that the birds do not die due to electrocution. Similarly for small cattle and large cattle, voltage can be determined to anesthetize the animal without arresting the heart and killing it.

Mushroom-Shaped Hammer Stunner

This is used in small slaughterhouses for large cattle and delivers a softer blow and stuns the animal for a short time. This method is not only acceptable but preferred over captive bold stunning in certain countries.

Carbon Dioxide Stunning or Gassing

Gassing is equivalent to chemical strangulation. Death by strangulation is prohibited; hence this method is not recommended.

HCP4

One of the requirements of halal slaughter is that the knife must be sharp so that the animal does not feel the pain of the cut. It is even more important for the knife to be sharp when the animals are slayed without any stunning. The sharp, swift single blow and gushing out of the blood triggers an anesthetic reaction in animals. The size of the knife should be proportioned to the size of the neck so that one may not have to use several back and forth strokes. The knife must not be sharpened in front of the animal.

HCP5

The slaughter person must be an adult male or female of sound mind familiar with the process of slaughter. He or she must not be weak at heart. A trained slaughter person will be more efficient and minimize damage to the skin and carcass.

HCP6: Slaying/Killing or Bleeding

The slaughter person must, while pronouncing the name of God and with a swift blow, cut the front part of the neck, severing carotids, jugulars, trachea, and esophagus, without reaching the bone in the neck.

HCP7: Invocation

It is mandatory to pronounce the name of God while cutting the throat. It suffices to say Bismillah (in the name of God) once; however, in general practice, especially for large animals, slaughter persons pronounce the name thrice as Bismillah Allahu Akbar, Bismillah Allahu Akbar.

HCP8: Postslaying Treatment

It is abominable to sever parts such as ears, horn, and legs before the animal is completely lifeless. Normally when the bleeding has ceased, the heart stops, and the animal is dead, one may start further acts of processing the carcass. Then, removing the skin and internals before deboning is carried out in a manner that protects the safety of meat.

HCP9: Packaging and Labeling

Packing is then done in clean packages and boxes, and proper labels are affixed to identify the products with halal markings.

HALAL REQUIREMENTS FOR BONING ROOMS

The same conditions that apply to abattoirs also apply to boning establishments, with respect to segregation of halal from non-halal, cleaning, certification, packing, and impressing cartooned products with a halal stamp. It is imperative that all halal products coming into the boning room from other establishments be accompanied by a halal transfer certificate and received by a Muslim inspector (Ayan, 2001).

HALAL REQUIREMENTS FOR COLD STORES

- All incoming halal load must be received by a Muslim inspector if the products are not sufficiently sealed or shipped in a bulk container.
- Halal products must be separately stored during blast freezing.
- Frozen halal products must remain isolated from non-halal products in the freezer unless packaged and sealed properly.
- Halal products should be loaded out separately from non-halal products under the supervision of a Muslim inspector. In a mixed halal and non-halal container, Halal products should be placed above non-halal products to avoid potential cross-contamination.
- All halal products transported out from the cold store must be accompanied by a transfer certificate for bulk-packed containers.
- All halal products loaded for export must be accompanied by a halal certificate.

FURTHER PROCESSED MEAT ITEMS

Meat and poultry products can be marketed fresh or frozen, and can also be used for further processing into processed meat items such as wieners, hot dogs, or chicken nuggets. A simplified flow chart (Figure 6.2) is given to point out HCPs in further processing:

HCP1: Meat Source

Halal inspector and quality assurance personnel make sure that meat received from a slaughterhouse is acceptable as halal to the halal-certifying agency. If hand-slaughtered poultry is specified, the company should not

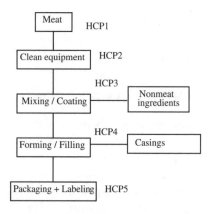

Figure 6.2 Further processed meat products.

use machine-slaughtered poultry. It is advisable to check with the halal agency before proceeding with the processing.

HCP2: Equipment

The equipment used for halal must be clean as inspected by the halal inspector as well as quality assurance and the department of agriculture. The company may use equipment that is used for non-halal meat and poultry after thorough cleaning, but not the equipment that has been used for pork processing. It is almost impossible to clean meat-processing equipment under normal operational conditions, which is why pork and nonpork equipment must be segregated. If one has to convert the equipment that has been used for pork to make halal products, the equipment must be ritually cleansed. An acceptable method includes the following: (1) Thorough washing of the equipment with hot water and detergent to get rid of visible traces of non-halal meat. (2) Rinsing of the equipment thoroughly with clean water by a Muslim inspector to make it acceptable for halal production. This is sometimes known as ritual cleansing.

HCP3: Nonmeat Ingredients

Thousands of ingredients are approved for use in meat products. One must ensure that prohibited material does not gain entry into halal products. Some of the ingredients to be avoided include, but are not limited to, gelatin, lard, pork extract, natural bacon flavor, other ingredients derived from animals, and ingredients containing more than 0.5% alcohol.

Appendix N lists halal, doubtful, and haram ingredients. Meat packers generally receive nonmeat ingredients from spice companies or directly from ingredient manufacturers. It is advisable to ask suppliers for halal certificates for all blended products or complex materials such as seasonings and spice blends; batters and breadings; and smoke and other flavorings.

HCP4: Casings

Casings can be edible or inedible. Some meat products are processed with the use of casings whereas others are not. Three types of casings are available and used according to the type of product:

- Natural casings — these are made from animal guts. These can be from lambs, sheep, goats, cows, or even pigs. Pig casings must not be used for halal products. Among other animals, casing can be from halal-slaughtered animals, generally from Muslim countries, or from non-halal-slaughtered animals, generally from the Western countries. It is recommended to use casings from halal-slaughtered animals.
- Collagen casings — these are made from finely ground cattle skins or theoretically can be made from pork skins. Because these are edible casings, they should be from halal-slaughtered animals.
- Cellulose casings — these are not edible casings. They are peeled off after the product is formed and cooked. Cellulose casings are made with cellulose (a plant material) and other ingredients such as glycerin. Halal-certified casings are available from major manufacturers.

HCP5: Packaging and Labeling

The final step in the manufacture of processed meat is packing the product in the right containers and labeling them to accurately identify them with halal markings. Examples of halal markings on product labels are given in Figure 6.3 and Figure 6.4.

Because the processing of meat and poultry varies a great deal from one company to the other and one type or species of animal to the other, the industry is advised to consult its halal certification authority for clarification of any issues.

INDUSTRIAL HALAL SLAUGHTER (DHABH) PROCEDURE

Under the Islamic jurisprudence (Shari'ah), a typical halal slaughter procedure is as follows. There are primary requirements that must be met, and there are secondary requirements that are not mandatory but merely recommendations.

GYROS CONES
MADE WITH ZABIHA HALAL MEAT

INGREDIENTS: BEEF, WATER, BREAD CRUMBS (FLOUR, WATER, SALT & YEAST. ALSO IT MAY CONTAIN PURE VEGETABLE SHORTENING, (SOYBEAN AND/OR COTTONSEED OIL), AMMONIUM CHLORIDE, CALCIUM PROPIONATE), SOY PROTEIN CONCENTRATE, SALT, MONOSODIUM GLUTAMATE, NATURAL SPICES, ONION, GARLIC.

OLYMPIA FOOD INDUSTRIES INC.
5757 WEST 59TH STREET
CHICAGO, ILLINOIS 60638

NET WT. 40 LBS

KEEP FROZEN
For Institutional
Use Only

Figure 6.3   Halal labeling of meat products: Gyros Cones product label. (Courtesy of Olympia Food Industries, Inc., Chicago, IL.)

Figure 6.4   Halal labeling of meat products: Beef Stew product label. (Courtesy of JRM Food Products Company, Deerfield, IL.)

Primary Requirements

- Animal or bird must be of halal species and alive at the time of slaughter.
- Slaughtering must be done by a mature Muslim of sound mind, trained in slaughtering method for the type and size of animal to be slaughtered.
- The name of Allah (Bismillah Allahu Akbar) must be verbally invoked by the Muslim slaughter person while slaying the animal.
- Slaughtering must be carried out on the neck from the front cutting the esophagus, wind pipe (trachea), jugular veins, and carotid arteries, without cutting the spinal cord beyond the neck muscle.
- Slaughtering must be carried out by a sharp knife in a swift sweep so that the animal does not feel the pain of a slaying.
- Blood must be drained out thoroughly and the animal must die of bleeding rather than any other injury, inflicted or accidental.

Secondary Requirements

- Animal or bird to be slaughtered should be healthy and free from diseases and defects.
- Animal or bird should be given water and handled humanely before slaughtering so that it is calm and not stressed out or excited.
- Slaughter person should be facing Mecca while slaughtering is carried out.
- Appropriate desensitizing or restraining method can be used to control the animal provided the animal is not dead before actual bleeding according to dhabha standards. If animal dies as a result of the desensitizing method, the animal carcass becomes prohibited (haram) for Muslim consumption.
- No part of the body should be cut prior to the actual slaughter or after slaughtering until the animal is completely dead.

Abominable Acts

It is not recommended to:

- Starve the animal by holding back food and water.
- Hold the animal down and then sharpen the knife.
- Sharpen the knife while the animal is looking at it and frighten it.
- Cut off the head or let the knife reach the bone.
- Break the neck of the animal while it is bleeding.
- Skin the animal while it is still alive.
- Use a dull knife to perform dhabha or use a knife of the wrong size.

INDUSTRIAL HALAL PROCEDURE FOR MECHANICAL SLAUGHTER OF POULTRY

■ The birds must be of halal species: chicken, ducks, or turkey.
■ The slaughter person while pronouncing Bismillah Allahu Akbar starts the machine.
■ The birds are hung onto the conveyor railing one at a time without agitating them.
■ The birds are passed over electrified water, touching the beak to shock them unconscious.
■ A Muslim slaughter person is positioned behind the machine and bleeds the birds missed by the machine while continuously invoking the name of God. (Two Muslim workers might be required, depending on the line speed.)
■ Halal birds are completely segregated throughout the process.
■ Containers of chilled birds are labeled halal.
■ The birds are cut up, deboned, and processed on thoroughly cleaned equipment.
■ Further processing such as marinating, breading, and packing off are done under the supervision of an inspector. Nonmeat ingredients must not contain any non-halal ingredients.
■ Products are properly marked with halal markings.

Industry Perspective on Halal Production

According to Jackson (2000), until now many Muslims accepted kosher meat products because they believed the slaughter was similar to their requirements and because the animals at least received a blessing at the time of slaughter. They are now learning that this is not true and are less accepting of kosher meat as a halal substitute. Until recently, commercial halal-certified meat products were virtually nonexistent in U.S. supermarkets except for imported products and locally slaughtered meat. Internationally, only proper halal certification is acceptable, and monitoring agencies are being established to enforce halal requirements. This international attitude is moving into the U.S. market (Jackson, 2000).

Some meat producers think that to be halal, they only had to follow a book of procedures. Companies following this policy will encounter marketing problems. The following are some corrections to notions on what can be considered halal meat (Jackson, 2000):

■ Muslim inspectors cannot say a blessing on a truck as it passes their houses on its way from the slaughterhouse to qualify the resultant meat as halal acceptable.

- Inspectors cannot say a blessing only at the start of the slaughtering process. It must be said throughout the process on each animal as it is slaughtered.
- A Muslim cannot say a blessing after all slaughtering is completed to cover all animals slaughtered that day.
- Inspectors cannot use recordings of blessing to substitute for the devotion of an observant Muslim.
- Meat producers cannot accept the word of the slaughterhouse that humane methods were used and the meat therefore should be considered halal.
- Producers cannot accept that a product labeled as halal is indeed produced halal. It must be certified or accepted by certifying agencies.
- Producers must never label a meat product as halal if there is no onsite Muslim participation. This is where the U.S. has lost competitively in the international market.
- Producers cannot simultaneously process any pork or pork-derived product while producing halal-labeled meat.
- Producers cannot process any pork or pork-derived product immediately prior to the processing of any halal-labeled meat product without a full, comprehensive, and detailed cleaning.

REFERENCES

Ayan, A.H. 2001. Halal food with specific reference to Australian export, *Food Aust.*, 53(11), 498-500.

Jackson, M. 2000. Getting religion — for your products, that is, *Food Technol.*, 54(7), 60-66.

Khan, G.M. 1991. *Al-Dhaba: Slaying Animals for Food the Islamic Way*, Abdul-Qasim Bookstore, Jeddah, Saudi Arabia.

Regenstein, J.M. and Chaudry, M. 2001. A brief introduction to some of the practical aspects of kosher and halal laws for the poultry industry, in *Poultry Meat Processing*, Sams, A.R., ed., CRC Press, Boca Raton, FL.

Regenstein, J.M. and Grandin, T. 2002. Animal welfare: kosher and halal, *IFT Relig. Ethn. Foods Div. Newsl.*, 5(1), 2.

Sams, A.R. 2001. *Poultry Meat Processing*, CRC Press, Boca Raton, FL.

7

HALAL PRODUCTION REQUIREMENTS FOR DAIRY PRODUCTS

The dairy products industry comprises a vast number of products ranging from fresh milk to ice creams and frozen desserts. Dairy is one of the oldest food industries. Cheese, which has been produced for about 5000 years, is one of the classical fabricated foods in the human diet (Fox et al., 2000). A wealth of information is available on all aspects of milk and dairy processing; however, the processors generally do not consider halal a part of their daily jargon. It would be appropriate here to mention that producing halal products is in essence similar to producing kosher products, although halal requirements are a bit simpler. Unlike meat products which can be either kosher or halal, but not both (Riaz, 2000a), dairy products can be kosher and halal at the same time. For a comparative analysis between kosher and halal, the reader is encouraged to consult Chapter 18.

There are two main dairy cattle in the world: the buffalo in South Asia and parts of Africa, and the cow in the rest of the world. There are other minor milk sources such as goats, sheep, mares, and camels. In this chapter, milk means cow's milk, although the processing of milk from other kinds of animals is similar to cow's milk.

MILK IN THE QURAN

Milk is one of the recommended foods for Muslims. It is considered pure and palatable for drinkers (Pickthall, 1994):

And lo! In the cattle there is a lesson for you. We give you to drink of that which is in their bellies, from between the refuse and the blood, pure milk palatable to the drinkers.

Chapter XVI, Verse 66

Milk is almost a complete food. It provides nourishment, minerals, vitamins, and protein. There are different types of dairy products from minimally processed such as fresh milk and cream to highly processed products such as sauces, dressings, and desserts. Accordingly, halal issues vary from simple to complex.

MILK: WHOLE, LOW-FAT, SKIM, AND FLAVORED

In the U.S., milk is generally fortified with Vitamins A and D. To make these vitamins soluble in milk, they are mixed with or standardized with emulsifiers such as polysorbates. Other functional ingredients can also be added to increase the stability and shelf life of milk. Polysorbates are fatty chemicals which can be made from vegetable oils or from animal fats. For producing halal milk, these emulsifiers and other functional ingredients must be from halal sources such as plant oils. Some dairies during chocolate and other flavored milk processing use gelatin as a thickener agent. To make it suitable for Muslim consumers, they should use halal gelatin or vegetable gums.

CREAM, HALF AND HALF, AND BUTTER

Mono- and diglycerides are sometimes added to these products to prevent the fat phase from separating from the water phase. Both animal- and vegetable-derived monoglycerides are available and only vegetable ingredients should be used for this purpose.

DRY MILK POWDER AND NONFAT DRY MILK POWDER

These are heat-processed, dehydrated milk powders. Normally no other ingredients are added to them. However, the spray drier must not have run any haram products.

CHEESES

There are many different types of cheeses, and they are processed by using different methods and different ingredients. Cottage cheese, for example, may be made by curdling milk with acid, which makes it a halal-suitable process. Other cheeses such as mozzarella, cheddar, and colby are made by the use of milk-curdling enzymes and bacterial cultures. Bacterial cultures are generally halal, as long as the media they are grown in are halal, but enzymes can come from many different sources, as explained in Chapter 12. One must make sure that enzymes are halal suitable. Some cheeses are ripened or aged by using bacterial cultures, molds, or enzymes. Cheeses processed with enzymes are more complex and might contain some objectionable ingredients. Transgenically produced enzymes are not only permitted but are preferred for use in the production of halal foods. For example, bovine rennet produced from calves that have not been slaughtered according to Muslim requirements is not acceptable by most Muslims, whereas chymosin (the main enzymes found in rennet) produced microbially through transcription from bovine chymosin genes is universally accepted by Muslims, as long as the standardizing ingredients and media in which they were raised contain no haram ingredients nor has the fermenter been used previously with haram ingredients without receiving a halal-approved cleaning. (*Note*: Because most of these products are almost always kosher, these requirements are routinely met. However, in the absence of a kosher certification, a more comprehensive halal inspection will be necessary.)

Enzymes are a major area of concern for halal cheese production. Several enzymes are obtained from pigs, which are haram. Some enzymes are also derived from calves or other permitted animals, but if these animals are not halal slaughtered the enzymes are not acceptable for halal cheese production. For enzyme manufacturers there is an opportunity to capture the halal market by making enzymes such as lipase through genetic engineering.

Anticaking Agents

Shredded cheese might contain anticaking agents such as animal or vegetable stearates. These ingredients should be from halal sources.

Preservatives

All preservatives and mold inhibitors should be from halal sources. The preservatives can be proprietary mixtures of natamycin, sodium benzoates, calcium propionates, and others, which may contain emulsifiers from animal sources, and so can be a concern for Muslim consumers.

The above ingredients, especially enzymes, not only affect the status of cheeses but also have a major impact on cheese by-products such as whey. Because whey contains a considerable amount of enzymes, Muslim consumers are concerned about foods containing whey (Riaz, 2000b).

WHEY, WHEY PROTEIN CONCENTRATE, WHEY PROTEIN ISOLATE, AND LACTOSE

These are by-products of the cheese-making process used in a myriad of food products ranging from baked goods to frozen desserts. These ingredients are further processed into powders from liquid whey. Usually, nothing further is added to whey after it is drained to make cheese; therefore, if the cheese manufactured is halal, then liquid whey and ingredients from whey, such as whey powder, whey protein concentrate, whey protein isolate, and lactose, are also halal, as long as the drying equipment is halal.

CULTURED MILK, SOUR CREAM, AND YOGURT

These are compounded, cultured, and further processed products of milk. Ingredients such as gelatin, emulsifiers, flavorings, stabilizers, and colors can be added to these products for various functional properties. Gelatin is one of the most widely used ingredients in yogurt. Halal gelatin is now available, as are other halal-texturizing ingredients, including pectin, carrageenan, and modified starches, which are suitable as gelatin replacements. Figure 7.1 shows a halal yogurt label.

ICE CREAMS AND FROZEN DESSERTS

Ice creams and frozen desserts are complex food systems requiring dozens of different ingredients to manufacture them. There are several possibilities for doubtful ingredients being incorporated into ice creams and frozen desserts, but the three ingredients that present the greatest difficulties are gelatin, flavors, and emulsifiers. To make natural vanilla ice cream, a company must use natural vanilla flavor, which by its "standard of identity" must contain at least 35% alcohol. Even when this alcohol is diluted down to its use level, the final ice cream may contain from 0.2 to 0.5% alcohol. Other flavors, which are uniquely liquor flavors, such as rum, might contain even higher amounts of alcohol. One must decrease the amount of alcohol in these products to the minimum level needed to provide the technical effects desired from that particular ingredient. The final alcohol content of the finished products should be lowered to less than 0.1%. Halal gelatin and halal marshmallows are now available for the companies that wish to formulate ice cream products with those ingredients. Flavors which have a connotation related to alcoholic beverage can be formulated from nonalcoholic extract of natural flavors or by the use of synthetic ingredients, such as rum flavor.

Figure 7.1 A halal frozen fruit bar label: Fruit*full* brand product label. (Courtesy of Happy and Healthy Products, Inc., Boca Raton, FL.)

FLAVORS AND ENZYME-MODIFIED PRODUCTS

These days many flavors such as butter flavor are extracted from dairy ingredients by concentrating the flavor compounds and further intensifying the flavor with modifications through enzymatic reactions. Manufacturers should make sure that enzymes used for these flavors are halal suitable. Currently, cheeses are becoming available where animal flesh ingredients are incorporated into the cheese. Such animal ingredients must be from halal source.

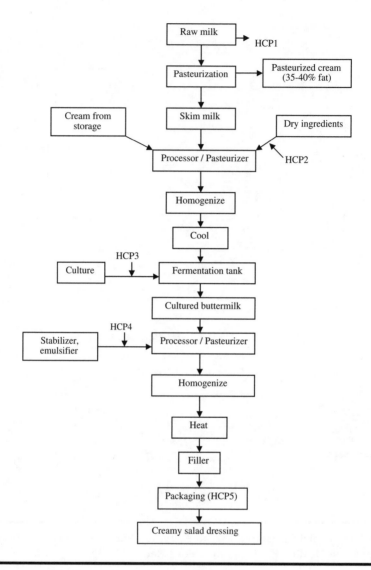

Figure 7.2   Halal control points for salad dressing.

HALAL CONTROL POINTS IN CHEESE MAKING

Cheese making generally is a dedicated process for each type of cheese, and there are very few chances of cross-contamination. It also is a fixed process, requiring few changes, if any, during the year. Figure 7.2 and Figure 7.3 give the halal control points (HCPs) for dairy technology.

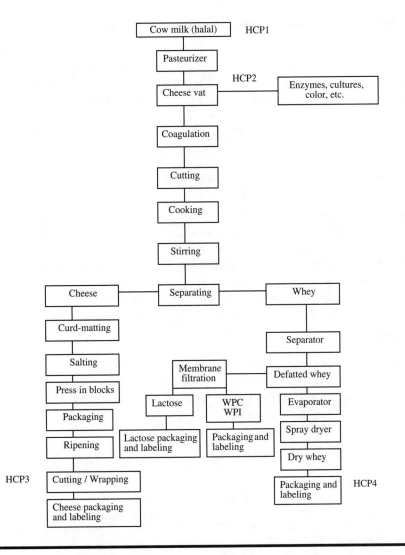

Figure 7.3 Halal control points for cheese and whey processing.

HCP1: Raw Milk

In the U.S., all milk is cow's milk unless specified otherwise; for example, cheese is also made from sheep and goat's milk. Other sources of milk may be used as long as they are halal.

HCP2: Additions of Enzymes, Cultures, and Colors

This is the most critical point in cheese manufacture because enzymes used can be from haram or halal animals as well as from microorganisms. For universal acceptance, enzymes should be of microbial origin and other ingredients should be free from doubtful ingredients. The use of these ingredients should be properly documented.

One must consider not only the ingredient itself but also other materials which might be added to the main ingredient to standardize or stabilize it.

HCP3

After the cheese is processed and ready to be stored for aging or ripening, mold and bacteria can be applied to its surface to be incorporated internally. Preservatives such as sorbates, propionate, or natamycin can also be applied where cheese is ripened without such cultures in order to inhibit mold growth. All these specialized chemicals must conform to halal guidelines.

HCP4

Finally, packaging must be done in clean halal suitable bags, wrappers, or boxes. If wax is applied as a moisture barrier or preservative, it must be of clean halal quality. Labels should have clear halal markings. It is advisable to use the description *microbial enzymes* if microbial enzymes are used.

REFERENCES

Fox, P.F., Guinee, T.P., Cogan, T.M., and McSweeney, P.L.H. 2000. *Fundamentals of Cheese Making*, Aspen Publishers, Gaithersburg, MD, p. ix.

Pickthall, M.M. 1994. Arabic text and English rendering of *The Glorious Quran,* Library of Islam, Kazi Publications, Chicago, IL.

Riaz, M.N. 2000a. What is halal (press article), *Dairy Foods*, 101(4), 36.

Riaz, M.N. 2000b. How cheese manufacturers can benefit from producing cheese for halal market. *Cheese Mark. News*, 20(18), 4, 12.

8

HALAL PRODUCTION REQUIREMENTS FOR FISH AND SEAFOOD

In this chapter, fish and seafood refer to all nonplant life from natural bodies of water, including rivers, lakes, ponds, seas, and oceans, as well as human-made fish farms. There are a number of differing opinions on the halal or haram status of fish and seafood. The Quran states (Pickthall, 1994):

> *To hunt and to eat the fish of the sea is made lawful for you, a provision for you and for seafarers; but to hunt on land is forbidden you so long as ye are on the pilgrimage. Be mindful of your duty to Allah, unto Whom ye will be gathered.*
>
> *Chapter V, Verse 96*

> *And He it is Who hath constrained the sea to be of service that ye eat fresh meat from thence, and bring forth from thence ornaments which ye wear. And thou seest the ships ploughing it that ye (mankind) may seek of His bounty, and that haply ye may give thanks.*
>
> *Chapter XVI, Verse 14*

> *And two seas are not alike; this, fresh, sweet, good to drink, this (other) bitter, salt. And from them both ye eat fresh meat and derive the ornament that ye wear. And thou seest the ship cleaving*

them with its prow that ye may seek of His bounty, and that haply ye may give thanks.

Chapter XXXV, Verse 12

These verses state that it is lawful to fish for food. In fact, God has subjected the seas to human beings so they may partake of their bounty and benefit by what has been provided. In addition, a number of ahadith (traditions of Muhammad) also address the subject of seafood (Al-Quaderi, 2002).

It is stated in a tradition of Muhammad that a group of his companions ran out of food on a journey and came upon a huge sea creature, often referred to as a huge fish or whale, washed up on the shore. They debated whether it was permissible to eat from it because it was already dead, but finally decided that their need for food exempted them should there be any sin in it. After returning home and informing Muhammad, they were told it was a blessing provided to them by God. Three points of jurisprudence have been established here:

- It is permissible to eat whale even though it is not considered a true fish because it is a mammal. Similarly, animals that wholly live in water (not water and land) are permitted for food.
- There is no requirement to slaughter sea animals similar to land animals, even if they are mammals. They do have to be killed humanely, generally by leaving them out of the water to let them die their natural death.
- Unlike land animals, it is permitted to eat dead sea animals. However, they must not show signs of deterioration and spoilage.

Islamic scholars have studied this question of which seafood is permitted and which is prohibited to be eaten by Muslims. Some of the scholars believe that only live catches are halal. They believe that if the object is found dead, it comes under the restriction of prohibiting the consumption of dead land animals. The majority of scholars opine that seafood is exempt from this restriction, and use the tradition about the dead whale to justify their opinion.

As to the species of sea creature that are permitted, all scholars have agreed that fish with scales are halal. Some believe that only fish with scales are halal and other creatures are not. This group believes that lobster, shrimp, octopus, eels, etc., are not permitted. Some have opined that anything that can only live in water is halal, whereas creatures that can live in and out of the water are haram. The latter include turtles, frogs, and alligators.

From this discussion, it seems that fish and seafood can be divided into four categories, with some categories universally accepted as halal, whereas others accepted by some people and not by others.

■ Category one — includes fish with scales and fins such as cod, flounder, haddock, halibut, herring, mackerel, perch, pollock, salmon, sea bass, whiting, buffalo fish, carp, trout, tuna, orange roughy, and snapper. This category is acceptable by all the Muslim consumers.

■ Category two — includes fish or fishlike animals which may have fins but not removable scales. Some of these may breathe oxygen from air rather than water, but live in water all the time. Examples are catfish, shark, swordfish, eel, monkfish, cusk, and blowfish. This category is acceptable to the majority of Muslim consumers, but not all denominations accept them as halal. They might consider them makrooh (disliked or detested).

■ Category three — comprises several unrelated species, mobile or not, of various shapes and sizes, that cannot survive without being in water. These are generally either molluscs or crustaceans, including clams, mussels, lobsters, shrimp, oysters, octopus, scallops, and squid. This group also includes marine mammals that live totally in the sea such as whales and dolphins. The majority of Muslim consumers eat them; however, others consider them either haram or makrooh. Shrimp seems to be in a special category: some only eat them but not the rest of category.

■ Category four — includes many of the animals generally falling under the definition of seafood. They live in and around water most of their life cycle, but are capable of living outside water because they can breathe air. These are generally not considered halal although some Islamic scholars are of the opinion that they are from the seas because they live in and around water. These include crabs, snails, turtles, alligators, and frogs.

REQUIREMENTS FOR SLAUGHTERING OR KILLING FISH AND SEAFOOD

Fish or any animals from the water are not required to be killed in any religiously specified manner as practiced for land animals. However, fish and seafood should be prepared in a manner that the animals do not suffer excessively. They should not be skinned or scaled while still alive, for example, as practiced by some Eastern countries.

GENERAL GUIDELINES FOR PROCESSING FISH AND SEAFOOD

General guidelines for processing fish and seafood consist of maintaining the identity of the product and not using any prohibited ingredients during processing. The guidelines also include not using equipment that has been used for haram products.

FURTHER PROCESSED PRODUCTS

Products such as fish fillets, fish sticks, fish patties, and other battered and breaded products must not contain any haram or doubtful ingredients, as outlined in Appendix N.

FOOD INGREDIENTS AND FLAVORS

It is common practice to extract flavors and manufacture ingredients from fish and seafood for use in nonseafood products. Because there is a large difference in the acceptability of fish and seafood among different Muslim consumers in various countries, it is advisable to learn the market requirements in a particular country before shipping any seafood products or flavors.

IMITATION SEAFOOD PRODUCTS

Some people do not consume real seafood, such as lobster tails and crab legs, either due to religious prohibitions or personal inhibitions, but eat imitation products. Imitation seafood products are generally made with surimi. Surimi is made from bland species of fish such as pollock. Imitation seafood products are flavored with natural and artificial flavors and texturized by using different stabilizing agents. The ingredients used in the preparation of imitation seafood products must conform to the general guidelines for halal products; that is, that they must not contain ingredients of haram animal origin or alcohol or any doubtful materials. Processing of surimi for manufacturing imitation seafood products is a unique process (Figure 8.1).

Halal Control Points in Surimi Production

The processing of surimi is rather simple. Animals and fish caught with target fish are removed, and rest of the fish is processed by washing, deheading, eviscerating, sometimes filleting, mincing, washing the minced meat several times, rewashing and squeezing the excess water and soluble proteins out, blending with stabilizers and cryoprotectants, and packing in boxes and freezing for further processing. There are only three halal control points (HCPs):

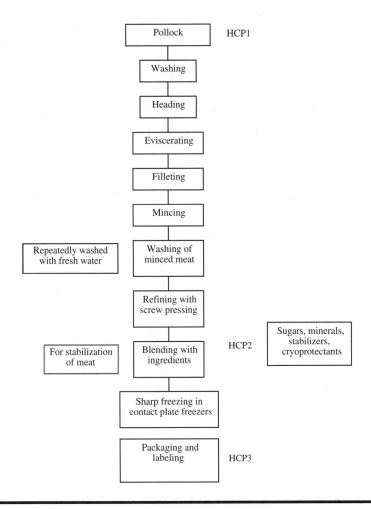

Figure 8.1   Processing of surimi made from pollock.

HCP1

This involves removal of nontargeted animals, which are removed anyway in the process because surimi cannot be made out of crab, shellfish, or turtles.

HCP2

This involves addition of stabilizers and cryoprotectants. Several ingredients such as sugar, sorbitol, calcium products, phosphates, egg whites, gelatin, blood plasma proteins, transglutaminase, or other binders and gums are used. If gelatin is used in surimi production, it should be halal bovine or fish gelatin. Similarly, any of the novel ingredients used in surimi production must also meet halal requirements.

HCP3

Finally, surimi should be packed in clean suitable halal packing materials and labeled properly and identified with halal markings.

REFERENCES

Al-Quaderi, S.J. 2002. Seafood: what's halal, what's not. *Halal Digest,* September, www.ifanca.org.

Pickthall, M.M. 1994. Arabic text and English rendering of *The Glorious Quran*, Library of Islam, Kazi Publications, Chicago, IL.

9

HALAL PRODUCTION REQUIREMENTS FOR CEREAL AND CONFECTIONARY

Cereal-based products include a large number of staple food products such as bread, breakfast cereals, cakes, candies, doughnuts, cookies, pastries, and chewing gum. The processing and composition of these products vary a great deal. Major ingredients used in this group of products are flour, sugar, and shortening. There are possibly hundreds of other ingredients used in cereal products, depending on the nature of each product (Riaz, 1996).

Some of the commonly used ingredients in cereal and confectionary industries whose halal status is questionable include the following.

- Gelatin — may be used as a glaze component on doughnuts and strudels as well as some types of cake and pastry. As discussed in Chapter 11, two types of gelatin are suitable for halal: (1) beef gelatin from halal-slaughtered animals, and (2) fish gelatin.
- Mono- and diglycerides — emulsifiers quite widely used in the bakery and confectionary industries, and, to a lesser extent, in candy products. Although mono- and diglycerides can be made from any fat and oil, the only acceptable sources for halal food are vegetable mono- and diglycerides (Riaz, 1998).
- Other emulsifiers — polysorbates of different molecular sizes, such as Tween 80, Tween 60, and Tween 40; sodium stearyl lactylate; and other specific-use emulsifiers are also questionable due to their sources. It is better to avoid emulsifiers from animal sources.

- Cream liquor — generally contains varying amounts of alcohol and must be avoided in halal production (Riaz, 1997).
- Pan grease and release agents — might contain ingredients such as wine, beef tallow, lard, gelatin, sugars, zein protein, or any other ingredient to create a coating of the food product. Although vegetables and mineral ingredients used for this purpose are halal, the coating formulators must avoid doubtful ingredients such as beef tallow and gelatin, or haram ingredients such as lard (Riaz, 1999). Sugars, zein, starches, bees wax, petroleum fractions, and vegetable oils are some of the halal-suitable ingredients for food coatings.
- L-Cysteine — amino acid that might be used in doughnuts, pizza crusts, taco shells, and tortillas. L-Cysteine might be used to modify the texture of the batters and breading. Halal L-cysteine must be either the vegetarian type made through synthesis or from bird feathers from birds slaughtered according to halal requirements. L-Cysteine from human hair is also available but generally not accepted as halal, because it is considered offensive to one's psyche, hence makrooh.

PRODUCT TYPES

Some of the key points in different product types such as breakfast cereals, bread, cookies and pastry, doughnuts and other fried goods, chewing gum, and marshmallows are discussed next.

Breakfast Cereals

Most of the breakfast cereals are rather simple formulations containing pure grain-based ingredients mixed in with sugar, salt, and a few other minor ingredients such as colors and flavors. The most widely used minor ingredients that are questionable include mono- and diglycerides and flavors. If there were no components in these animal-based ingredients, all cereals would be halal. However, some of the breakfast cereals may contain gelatin-based marshmallows which may not be halal (Sakr, 1999).

Bread

There is a large variety of breads, from flat, unrolled, made without yeast to flat, leavened, made with yeast, such as pita bread to French- and Italian-type rolls. Breads are also highly leavened and yeast raised, such as sliced breads. The earlier examples have simpler formulations. The latter ones such as white bread might contain many ingredients. Also, all breads are primarily made of flour and water, which might contain several minor ingredients. Some of the questionable ingredients in breadmaking are mono- and diglycerides, sodium stearyl lactylates, and flavorings. Another major concern in breadmaking is pan grease and release agents used in the utensils.

Many breads are made with yeast, which not only generates carbon dioxide but also produces alcohol. Although alcohol is one of the haram ingredients, the purpose in making the bread is not similar to brewing alcoholic drinks. Hence, there is no concern with the presence of any residual alcohol in bread.

Cakes, Cookies, and Pastries

The major components of these food products are flour, sugar, and oil. However, many of the minor ingredients might be used to give these products a specific character, which differentiates one type of product from the other. The minor ingredients to avoid in these products are mono- and diglycerides, gelatin, polysorbate, sodium stearyl lactylate, l-cysteine, flavors containing alcohol, and other non-halal ingredients.

Doughnuts and Other Fried Goods

The first and foremost requirement is that the frying oil should be vegetable oil or from halal source. The minor ingredients to avoid in doughnuts are very similar to those present in cakes and cookies. Gelatin might be used on glazed doughnuts, which can be easily replaced by an appropriate plant gum.

Chewing Gum

Gum-base and chewing gum might be composed of any of the generally recognized as safe (GRAS) by the USFDA ingredients. Two of the commonly used ingredients which are doubtful as regards halal include stearates and gelatin (Uddin, 1994). There might be many other doubtful ingredients in gum base that need to be screened properly for the production of halal chewing gum. Only halal gelatin and halal stearates are acceptable for halal products.

Marshmallows

Marshmallows are primarily sugar, gelatin, and flavors. A special kind of gelatin, usually pork, is required to make the best fluffy, nonsticky marshmallows. One can make halal marshmallows with halal-certified bovine or fish gelatin. Flavors used in marshmallows must also be halal suitable. During production, marshmallows in most plants are rolled in corn starch so that they do not stick to each other. It is very important that the equipment be thoroughly cleaned and old starch replaced with fresh starch before producing halal marshmallows. The equipment at the time of start-up must be physically inspected to remove any stray marshmallows from the previous batch.

HALAL CONTROL POINTS (HCPs) IN BREADMAKING

Although the process of breadmaking is quite simple, there are a few points to consider in the production of bread (Figure 9.1):

- HCP1 — ingredients used in the manufacture of bread. All major and minor ingredients must be halal suitable.
- HCP2 — release agents and pan grease if used should also be halal suitable.
- HCP3 — packaging materials used for packing halal bread must not contain any ingredients of animal origin such as animal stearate.

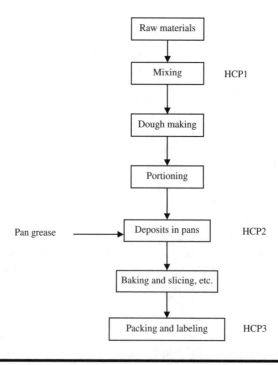

Figure 9.1   Flow chart of breadmaking showing halal control points.

REFERENCES

Riaz, M.N. 1996. Hailing halal, *Prep. Foods*, 165(12), 53-54.

Riaz, M.N. 1997. Alcohol: the myths and realties, in *A Handbook of Halaal and Haraam Products*, Vol. 2, Uddin, Z., Ed., Center for American Muslim Research and Information, Richmond Hill, NY.

Riaz, M.N. 1998. Halal food: an insight into a growing food industry segment, *Food Market. Technol.*, 12(6), 6-9.

Riaz, M.N. 1999. Examining the halal market, *Prep. Foods*, 168(10), 81-85.

Sakr, A.H. 1999. *Gelatin*, Foundation for Islamic Knowledge (Lombard, IL) and Islamic Food and Nutrition Council of America (Chicago, IL), pp. 13-28.

Uddin, Z. 1994. *A Handbook of Halaal and Haraam Products*, Publication Center for American Muslim Research and Information, Richmond Hill, NY.

10

HALAL PRODUCTION REQUIREMENTS FOR NUTRITIONAL FOOD SUPPLEMENTS

This chapter deals with dietary and nutritional supplements containing nutraceuticals, vitamins, minerals, and other nutritional products used to enhance and maintain health and wellness or prevent disease. In the past several decades, the number of supplements available to the consumers through specialty stores, supermarkets, and especially through multilevel marketing has seen a tremendous growth worldwide. The lines between pharmaceuticals, products that heal, and nutraceuticals, products that help maintain the well-being of a person, are merging. The Food and Drug Administration is taking a closer look at several dietary products for their health claims. The purpose of this chapter is not to determine the effectiveness of these products but to reflect on their compositions and determine whether any of the components presents a problem for the Muslim consumer.

Although in the Islamic tradition one may consume a haram product as a medicine under compulsion, Muslim consumers generally avoid knowingly taking anything which is haram or doubtful. Some people may take a prescription medicine in a gelatin capsule but not a multivitamin capsule. Gelatin capsules, unless certified halal or labeled bovine, are generally made of pork gelatin. Pork gelatin is considered haram by Muslim consumers. Medicine that is used to cure disease and help overcome illness is considered exempt from halal food regulations. Prescription drugs generally do not have alternative products to replace a prescribed drug. If a drug is available in capsule form only, one is obligated to take it, whereas multivitamins are normally not taken to cure serious illness, but to improve one's health. Moreover, there are many alternative forms of multivitamins, such as tablets, liquids, and vegetable capsules, so one does not have to take vitamins in gelatin capsules. Many consumers try to purchase alcohol-free products such as cough syrups. They can also ask the pharmacist for tablets rather than gelatin capsules. Malaysia is one of the most active markets for nutritional supplements and is a place where people look for halal versions of these products.

The Malaysian Department of Health, however, has determined that nutritional supplements are health and medication products and may not be grouped together with food products. It is against the local regulations to display halal markings on such products. Because Muslim consumers are very apprehensive of this regulation, the government might have to yield to the wishes of the consumers and repeal this ruling. Indonesian authorities, on the other hand, are including not only the foods but also drugs and cosmetics in their halal program, which is evident by the creation of an institute, the Assessment Institute for Foods, Drugs, and Cosmetics (AIFDC), under the guidance of the Religious Council of Indonesia, also called the Majelis Ulama Indonesia (MUI). AIFDC is responsible for assessing, evaluating, certifying, and monitoring establishments and products including foods, drugs, cosmetics, personal care products, and other consumables. AIFDC insists on using logos or proper halal markings on labels of certified halal products or ingredients. It is therefore important to determine the market acceptability before launching a product in any country or region.

General guidelines for the production of nutritional supplements are similar to producing other food products. Nutritional food supplements, for the most part, are composed of botanicals and plant extracts. It is the animal-derived ingredients one has to avoid in formulating the supplements. The botanical ingredients have been used in various cultures and traditions for centuries, such as ginseng in the Chinese culture (Li and Wang, 1998), black seed in Islamic tradition (Al-Akili, 1994), and asphoetida in India (Raghavan Uhl, 2000).

INGREDIENTS TO WATCH

Databases of ingredients can often comprise thousands of entries. Companies might use several thousand different ingredients in a given time period. It is beyond the scope of this book to describe the halal status of every ingredient used in the industry. Only some of the ingredients with potential concern for halal and the type of products they may be used in are given here.

- Flavors and colorants — might have hidden alcohol or ingredients of haram animal origin, such as civet oil, in the formulations.
- β-Carotene — often formulated with gelatin in small quantities. Gelatin is used to encapsulate and protect its color and other characteristics. Some companies use fish gelatin for encapsulation, which makes the product halal as well kosher. Manufacturers also use halal bovine gelatin or plant gums to encapsulate β-carotene.
- Gelatin — very commonly used to make capsules, both softgel and two-piece hard shell. Halal gelatin or cellulose or starch can be used instead of porcine gelatin.
- Stearates from animal sources — can be used as free-flow agents in powders or tableting aids in tablets. For halal products, manufacturers can use stearates from plant sources.
- Tweens — sometimes used for coating and polishing tablets. Vegetable-derived tweens rather than animal-derived ones should be used in halal products.
- Glycerin — used in the manufacture of capsules, and it may also be used in other products. Glycerin of plant origin is halal suitable for such applications.

TYPES OF PRODUCTS

Nutritional food products come in many physical forms such as powders, liquids, tablets, one-piece capsules (soft shell), and two-piece capsules (hard shell). Nutraceutical ingredients can also be incorporated into food matrices such as juices, snack bars, or energy drinks.

- Tablets — can be coated with gelatin (gel tabs) or with specialty lipids such as polysorbates. Halal-certified gelatin and lipids of plant origin should be used for halal tablets. Sugars and plant proteins such as zein can also be used as coating material for tablets in halal products.
- Liquid supplements and drinks — many liquid formulations are standardized with ethyl alcohol as a preservative or solvent. Alternatives such as mixtures of propylene glycol and water can be used. The amount of alcohol in the finished product may not be more than 0.1% as discussed in Chapter 13.
- Softgel capsules — one-piece capsules used to be made exclusively from gelatin. They can also be made with vegetable ingredients such as modified starch, cellulose gum, and other plant gums. Halal-certified bovine and fish gelatin is also available for this purpose. Besides the main ingredient, softgel capsules might also

contain glycerin or fatty chemicals, which should be from plant sources for halal production.

■ Hardgel capsules — like softgels, two-piece hardgel capsules used to be made exclusively with gelatin. There is now the advent of vegetarian capsules especially for nutritional supplements. Vegetarian capsules can be made from modified cellulose, modified starches, or other plant materials. Glycerin and other ingredients can be used as processing aids. All such ingredients should be from vegetable or petroleum sources. Two-piece gelatin capsules, if used, should be halal-certified bovine type and other incidental ingredients should also be halal suitable.

To ensure almost universal acceptability, it is recommended that pharmaceutical products and nutritional and dietary food supplements be manufactured by avoiding all traces of animal products, so that the product is acceptable for halal, kosher, and vegetarians or vegans.

REFERENCES

Al-Akili, M. 1994. Black seed, in *Natural Healing with the Medicine of the Prophet*, Pearl Publishing House, Philadelphia, PA, pp. 229-232 (translated and emended by Muhammad Al-Akili from *Book of the Provisions of the Hereafter* by Imam Ibn Qayyim Al-Jawziyya).

Li, T.S.C. and Wang, L.C.H. 1998. Physiological components and health effect of ginseng, echinacea, and sea buckthron, in *Functional Foods: Biochemical and Processing Aspects*. Mazza, G., Ed., Technomic Publishing, Lancaster, PA, p. 329.

Raghavan Uhl, S. 2000. A to Z spices, in *Handbook of Spices, Seasoning, and Flavorings*, Technomic Publishing, Lancaster, PA, p. 65.

11

GELATIN IN HALAL FOOD PRODUCTION

Gelatin is used in many food products, including jellies, ice cream, confectionery, cookies, and cakes. It is also used in nonfood products, including medical products, and in veterinary applications. Gelatin can be from halal or haram sources. Common sources of gelatin are pigskin, cattle hides, cattle bones, and, less frequently, fish skins and poultry skins. In general, a product label does not indicate the source of the gelatin (Chaudry, 1994), so halal consumers normally avoid products containing gelatin unless they are certified halal. As Muslim countries have increased imports of food products, there has been growing awareness of the problem gelatin presents to Muslim consumers. Malaysia, Indonesia, and several other Muslim countries now require that imported as well domestic products containing gelatin be produced with halal gelatin. Several gelatin manufacturers in Europe, India, and Pakistan produce halal gelatin.

STATUS OF GELATIN IN ISLAM

Gelatin is an animal by-product, the partially hydrolyzed collagen tissue of various animal parts. Its halal status depends on the nature of raw materials used in its manufacture. Most gelatin is one of two types: (1) Type A gelatin is exclusively made from pork skins, and is hence haram for Muslims to use. (2) Type B gelatin is made either from cattle and calf skins or from demineralized cattle bones. Cattle and calf skins used in gelatin manufacture are usually from animals slaughtered by non-Muslims. Whether this type of gelatin is permitted or prohibited for Muslims is controversial. However, gelatin made from bones of duly halal-slaughtered cattle is available. Fish-skin gelatin is halal as long as it is free from contamination from other sources and is made from a fish species accepted by Muslims who use the product. A food processor understands that a nonspecific gelatin is highly questionable regarding its source, highly suspected of containing pork gelatin, and very strongly discouraged for use by the Muslims (Sakr, 1999).

SOURCES OF GELATIN

For gelatin from cattle skins, cattle bones, poultry skins, or other permitted animals to be halal, the animals have to be slaughtered according to Islamic rites, as explained in Chapter 3.

In modern slaughterhouses, bones are sold to rendering companies, which turn fresh bones into dry bone chip used for gelatin. However, in many Asian and African countries, bones are discarded as waste and are subject to a natural degreasing process. Without going into details of the degreasing process, it suffices to say that when collecting and selecting the bones for food-grade or pharmaceutical-grade halal gelatin, bones must be examined and segregated into bones from halal species and those from non-halal species, which cannot be used. Bones from animals that have died without being properly slaughtered or that were used for religious ceremonies are also prohibited.

PRODUCTION OF HALAL GELATIN

Gelatin is derived from collagen, an insoluble fibrous protein that occurs in vertebrates and is the principal constituent of connective tissues and bones. Gelatin is recovered from collagen by hydrolysis. There are several varieties of gelatin, the composition of which depends on the source of collagen and the hydrolytic treatment used. Some typical gelatin production processes are shown in Figure 11.1.

Preparation of Source Material

The principal raw materials used in halal gelatin production currently are cattle bones and cattle hides. Noncollagen substances such as minerals (in the case of bone) and fats and proteins (in the case of hide) are removed by various treatments to prepare collagen for extraction.

■ Bones — fresh bones, also called green bones, from the halal-slaughtered cattle are cleaned, degreased, dried, sorted, and crushed to a particle size of ca. 1 to 2 cm. The pieces of bone are then treated with dilute hydrochloric acid to remove mineral salts. The resulting sponge-like material is called ossein.

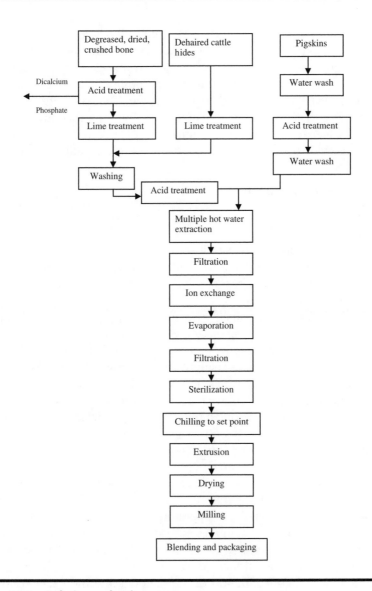

Figure 11.1 Gelatin production process.

■ Hides — cattle hides from halal-slaughtered animals are received from the trimming operations of leather production. The hide pieces are usually dehaired chemically with a lime and sulfide solution, followed by a mechanical loosening.

For the production of halal gelatin, both ossein and cattle hide pieces are subjected to lengthy treatment with an alkali, usually lime and water, at ambient temperature. Depending on previous treatment, nature of the material, size of the pieces, and exact temperature, liming usually takes 8 to 12 weeks. The process is controlled by the degree of alkalinity of the lime liquor as determined by titration with acid, or by making test extractions. Ossein usually requires more liming time than cattle hide does. Additional lime is added to maintain an excess, thereby compensating for any that is consumed. The material is then thoroughly washed with cold water to remove excess lime, its pH adjusted with acid, and it is extracted with hot water to recover the soluble gelatin.

Extraction of Halal Gelatin

The number of extractions typically varies from three to six. The first extraction generally takes place at 50 to 60°C; subsequent extractions are made with successive increases in temperature of 5 to 10°C. The final extraction is carried out close to the boiling point. Extracts are kept separate, analyzed, and subsequently blended to meet various customer specifications.

The initial extraction usually provides a superior product compared with subsequent extractions. Earlier extractions have higher molecular weight, higher viscosity, higher gel strength, and the least color. Later extractions are made at increasingly higher temperatures; the resulting product has lower molecular weight, lower gel strength, and greater color.

Dilute halal gelatin solutions from the various hot water extractions are filtered, deionized, and concentrated by cross-flow membrane filtration and vacuum evaporation, or both. Halal gelatin solution is then chilled and either cut into ribbons or extruded as noodles, and the gelled material is deposited as a bed onto an endless, open-weave stainless steel belt. The belt is passed through a drying chamber, which is divided into zones in each of which the temperatures and humidity of the drying air is accurately controlled. Typical temperatures range from about 30°C in the initial zone up to about 70°C in the final zone. Drying involves progressive increases in air temperature, often with exhaustion of moist air and replenishment with conditioned air. Drying time is 1 to 5 h, depending on the quality and concentration of the material and the exact conditions employed. The rate of drying is carefully controlled to avoid melting and case hardening. Halal gelatin leaves the dryer with moisture content of about 10%. The dried bed is then broken into pieces that are ground to the required particle size.

Where halal gelatin is not available, food manufacturers can use some of the vegetable substitutes for gelatin. These vegetable-origin substitutes perform the same function as gelatin. However, currently gelatin is the only material that melts below body temperature and is reversible, that is, can be melted and gelled more than once.

VEGETABLE SUBSTITUTES FOR GELATIN

- Agar — also called agar-agar, gelose, Chinese isinglass, Japanese isinglass, Bengal isinglass, or Ceylon isinglass.
- Carrageenan — polysaccharide extracted from red seaweed.
- Pectin — polysaccharide substance present in cell walls of all plants.
- Xanthan gum — polysaccharide gum produced by bacteria. The bacterial medium must be halal for the product to be halal.
- Modified corn starch.
- Cellulose gum.

USES OF HALAL GELATIN

Gelatin is a multifunctional ingredient widely used in food systems, pharmaceutical preparations, and other industrial products.

Gelatin in Foods

Halal gelatin can be used for gelatin dessert; dairy products such as yogurt, sour cream, and cottage cheese; and other dairy and imitation dairy foods. It is also widely used in frozen desserts such as ice creams, cream pies, and cheesecakes. Gelatin is the primary ingredient in marshmallows, keeping their whipped foam structure intact. It has been used in confections. In the meat industry, gelatin is used in luncheon meats, jellied beef, and corned beef loaves. Gelatin is also used as a processing aid in the food industry for the clarification of cider and fruit juices.

Halal Gelatin in Pharmaceuticals and Cosmetics

The major use of halal gelatin in the pharmaceutical industry is in the manufacture of capsules. Both soft and two-piece hard capsules as well as enteric capsules contain gelatin as the main ingredient. Figure 11.2 and Figure 11.3 show capsules made with halal gelatin and halal gelatin dessert, respectively.

Halal gelatin is also used in tablets as a binding, moisturizing, and coating agent. Lozenges and cough drops are often made with gelatin as the preferred base due to its lubricating and nonirritating properties. Gelatin has been recognized as an excellent stabilizer and emulsifier in pharmaceutical emulsions. It has been used for the external application of drugs to treat various skin disorders and as an adhesive to hold bandages and dressings together. Other pharmaceutical uses of gelatin include a glycerinated base for suppositories and a carrier for certain dietary

Figure 11.2 Halal gelatin capsule. (From an advertisement in the *Halal Consumer* magazine, issue 5, 2003, Chicago, IL.)

supplements. Gelatin is commonly used in various cosmetic preparations as an emulsifying and smoothing agent. It is frequently used in creams and wave-set lotions, and is the protein used in "protein" shampoos and hair conditioners. The use of halal gelatin in any of these products will increase the market for these products in Muslim countries.

Medicinal, Dietetic, and Therapeutic Uses

Gelatin in various forms is regularly used in the medical profession as absorbable sponges to arrest hemorrhage and as dusting powder for surgical gloves. Because gelatin is compatible with body processes, it does not give any complications in open wounds. Gelatin is an excellent dietetic and therapeutic agent for the prevention of obesity. Low-sugar gelatin desserts require more calories to digest than they contain. Gelatin (when properly supplemented) has been used as a protein food in malnutrition and infant feeding. Other therapeutic uses include treatment of digestive disorders, peptic ulcers, muscular disorders, and brittle fingernails. Gelatin is commonly used as a plasma extender for the treatment of shock. Pharmaceutical and drug industries should start using halal gelatin rather than just gelatin to cater to the Muslim market. Halal gelatin, which has the same functional properties as regular gelatin and might only be slightly more expensive, opens up many new markets.

(Front panel)

Nutrition Facts

Serv. Size 1/4 of Package (21g of mix)
Servings Per Container 4

Amount per serving

Calories 80	Calories from Fat 0

	% Daily Value*
Total Fat 0g	0%
Sodium 80mg	3%
Total Carbohydrate 19g	6%
Sugars 19g	
Protein 1g	0%

Not a significant source of Saturated Fat, Cholesterol, Dietary Fiber, Vitamin A, Vitamin C, Calcium and Iron.

*Percent Daily Values are based on a 2,000 calorie diet.

INGREDIENTS: Sugar, Halal Gelatin, Adipic Acid (for tartness), Contains less than 2% of Sodium Citrate (controls acidity), Fumaric Acid (for tartness), Salt, Artificial Flavor, Red #40, Dimethylpolysiloxane (prevents foam).

Distributed by:
USIF, Inc.
PO Box 18374
Chicago, IL 60618

DIRECTIONS: Add 1 cup (8 oz.) boiling water to the contents of this package. Stir until completely dissolved. Add 1 cup cold water. Chill until set.
TO ADD FRUITS OR VEGETABLES: Chill until slightly thickened, then add 1 to 2 cups cooked or raw fruits or vegetables. (Do not use fresh or frozen pineapple, kiwi, gingerroot, papaya, figs, or guava; gelatin will not set.) For molds reduce water to 3/4 cup.
SPEED SET: (Soft-set and ready to eat in 30 minutes.) Add 1 cup boiling water to contents of this package. Stir to dissolve completely. Add 1/2 tray (6 to 8) ice cubes, stir until slightly thick, remove unmelted ice.

(Back panel)

Figure 11.3   Halal gelatin dessert product label (front and back panels). (Courtesy of USIF, Inc. Chicago, IL.)

CONTROL POINTS IN HALAL GELATIN PRODUCTION

The most important issue in producing halal gelatin is obtaining the proper raw materials. Because pigskins cannot be used for halal gelatin production, plants that manufacture pigskin gelatin should not be considered for halal production. Halal raw materials, both bones and hides, are a limiting factor; hence, gelatin companies have to work upstream with their suppliers through tanneries and bone mills, all the way to halal slaughterhouses.

HCP1: Raw Materials

All sources, hides, and bone chips should be approved and constantly monitored. Gelatin factories normally receive pieces of hide and bone chips. Gelatin manufacturers must execute controls at their supplier's plants to make sure raw materials are properly segregated.

HCP2: Degreasing of Fresh Bones

This step is generally performed in the rendering plants, but halal gelatin manufacturers have to make sure that their suppliers have proper controls when degreasing the halal bone in order to minimize cross-contamination from non-halal sources.

HCP3: Acid Treatment

This step takes place in vats or pits. It is best if vats are dedicated for halal bones.

HCP4: Lime Treatment

A similar setup may be used for lime treatment. Again, it is advisable to use dedicated setup.

HCP5: Extraction, Evaporation, Extrusion, and Drying

These are generally carried out in tandem and continuously. The system and equipment should be thoroughly cleaned and the cleaning documented before starting a halal run.

HCP6: Milling and Blending

Gelatin is almost always ground to meet the granulation specification and gelatins of different bloom strengths are blended together to get the desired

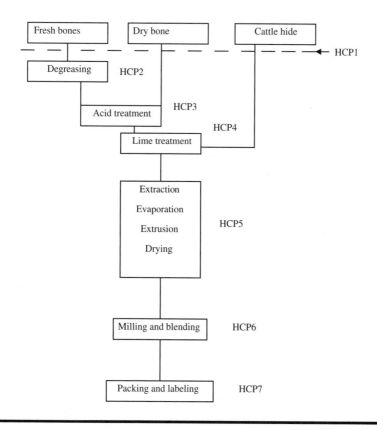

Figure 11.4　Gelatin production showing halal critical points.

gel strength. The mills and blenders including all charging, discharging, and conveying equipment should be properly cleaned to rid the equipment of any non-halal materials that were previously handled on the equipment.

HCP7

Finally, halal gelatin must be packed in clean containers and labeled properly with halal markings to avoid any mix-up with non-halal gelatin. Figure 11.4 shows all the HCPs.

REFERENCES

Chaudry, M.M. 1994. Is kosher gelatin really halal? *Islam. Perspect.*, XI(1), 6.
Sakr, A.H. 1999. *Gelatin*, Foundation for Islamic Knowledge (Lombard, IL) and Islamic Food and Nutrition Council of America (Chicago, IL), pp. 13-28.

12

ENZYMES IN HALAL FOOD PRODUCTION

Enzymes play a vital function in the regulation and performance of living cells. They speed up reactions and catalyze specific reactions while producing few by-products. The food industry has taken advantage of these properties to produce enzymes that can reduce food production costs, manufacturing time, and waste production, and improve taste, color, and texture. The advent of bioengineering has enabled the enzyme industry to produce microbial enzymes that look and function like animal or vegetable enzymes. Bioengineered enzymes are cheaper to produce and easier to control for purity. The use of microbial enzymes is more compatible with halal food production because it eliminates the use of animal-derived enzymes. However, animal-derived enzymes are still used in the food industry, particularly in the dairy industry.

USES OF ENZYMES IN FOOD

Enzymes have been used for many centuries. There is evidence of the use of enzymes in cheese making as mentioned in Greek epic poems dating back to 800 B.C. (Ashie, 2003). Today, enzymes are used for many purposes and in many industries. Major food industry uses are cheese making, baking, fruit and vegetable processing, and food ingredient production. Enzymes are used in production of sugar, processing of starch, hydrolysis of proteins, and modification of oil and fat. Other food uses include brewing and winemaking.

The first major food industry use of enzymes began in the 1960s, when the enzyme glucoamylase was developed to allow starches to break down to glucose (Olsen, 2000a). Until then, glucose production involved acid hydrolysis of starch. While the acid hydrolysis method produces the desired product, use of the enzymatic process reduces production costs, waste, and undesirable by-products. Today, almost all glucose is produced by using enzymes. High-fructose corn syrup, which is found in most soft drinks, is also produced by using an enzyme.

In the baking industry, enzymes help improve the quality, freshness, and shelf life of breads and baked goods. They convert sugars to alcohol and carbon dioxide, causing the dough to rise; strengthen the gluten network, resulting in greater flexibility and machinability; and modify triglycerides, resulting in larger loaf volumes.

In cheese making, enzymes help the milk to coagulate, the first step in making cheese. In the dairy industry, both microbial and animal enzymes are used. Enzymes are also used to accelerate cheese ripening and to reduce the allergic properties of dairy products. Chymosin is the enzyme used for coagulation, lipase is used for ripening, and lactase is used to improve digestibility. The source of animal enzymes is a concern for halal consumers. Cheese and whey produced by using animal enzymes are haram if the source is haram animals. Cheese and whey produced by animal enzymes from halal animals not slaughtered according to Islamic requirements are doubtful, because only some consumers accept these products. Because whey is found in many nondairy products, the use of animal-derived enzymes to produce whey poses a problem for the halal consumer (Chaudry, 2002).

Protein hydrolysis is another application for enzymes. Animal and vegetable proteins are hydrolyzed by using enzymes to improve functionality and nutritional values of proteins. Proteins are used in foods as emulsifiers; hydration agents; to control viscosity; as gelling agents; and to improve cohesion, texture, and solubility. Production of these proteins by chemical reactions is undesirable because it requires severe conditions and produces many by-products, which are difficult or expensive to remove. Enzymes are faster, allow production under milder conditions, and produce fewer by-products. Enzymes can be tailored to catalyze a specific reaction, with very low by-product formation. Enzymes can be used to produce meat extracts from scrap-bone residues. These are used in soups, sauces, broths, and other applications.

In the juice industry, enzymes increase yields and improve color and aroma. Enzymes are used to clarify juices and extract essential oils from citrus peels. Enzymes can also improve the texture of fruit pieces used in food products, such as fruit-flavored yogurt.

Enzymes can be used in the extraction of vegetable oil. Conventionally, oily materials such as rapeseeds, coconuts, sunflower seeds, palm kernels, and olives are first pressed to extract the oil. Remaining oil is then extracted by using an organic solvent. In this second phase, enzymes can be used to allow the oil to be extracted into a water solution, avoiding the need for an organic solvent, making it a more environmentally friendly method of producing oils. However, this method is not yet in wide use. Enzymes are also used to modify oils to improve nutritional value or to produce lubricants and cosmetic ingredients.

CLASSIFICATION OF ENZYMES

The International Union of Biochemists has developed an identification system to classify enzymes. Naming of enzymes involves a numerical classification, a long systematic name, and a short, easy-to-use name. For example, the enzyme that catalyzes the conversion of lactose (milk sugar) to galactose and glucose is classified as EC 3.2.1.23, has the systematic name β-ᴅ-galactoside galactohydrolase, and the common name lactase.

Enzymes are classified into six categories (Olsen, 2000b):

- Oxidoreductases — catalyze oxidation reactions, such as the conversion of alcohol.
- Transferases — enzymes that catalyze the transfer of a group of atoms, referred to as a radical, from one molecule to another, such as the transfer of amino groups.
- Hydrolases — catalyze the reaction of a chemical with water. This usually breaks up large molecules into smaller ones, such as the hydrolysis of proteins.
- Lyases — catalyze reactions producing or resulting in double bonds, for example, conversion of sugars.
- Isomerases — catalyze the transfer of groups on the same molecule, resulting in a new structure for the molecule.
- Ligases — enzymes catalyze the joining of molecules to form larger molecules.

BIOENGINEERED ENZYMES

The identification and understanding of DNA led to gene splicing and the development of bioengineered enzymes. Many enzymes are now produced by bioengineering, with a variety of methods employed for growth and production of enzymes. One method of production is submerged fermentation. In this process, selected microorganisms (bacteria and fungi) are grown in closed vessels in the presence of liquid nutrients and oxygen. As the microorganisms break down the nutrients, they release enzymes into the solution. The nutrients are normally sterilized foodstuff such as cornstarch, sugars, and soybean grits. The process can

Table 12.1   A Sampling of Major Sources of Enzymes

Source	Type	Activity	Uses
Bacteria	Bacillus	Protease	Meat, beverages
Bacteria	Streptomyces	Isomerase	Beverages, starch
Fungi	Aspergillus	Protease	Cheese
Fungi	Mucor	Lipase	Cheese, fat
Yeast	Saccharomyces	Invertase	Cocoa
Yeast	Kluyveromyces	Chymosin/rennin	Cheese
Plant	Barley/malt	Amylase	Bakery, sugar
Plant	Papaya	Papain/protease	Bakery, beverages
Animal	Bovine liver[a]	Catalase	Beverages, dairy
Animal	Ruminants[a]	Rennin/protease	Cheese
Animal	Pig/cattle stomach[a]	Pepsin/protease	Cheese, cereals

[a] Examples of enzymes that are of particular concern in halal food production.

Source: From Mathewson, P.R. 1998. Major biological sources of enzymes (Appendix C), in *Enzymes*, Eagan Press, St. Paul, MN, pp. 93-95. With permission.

be operated continuously or in a batch mode. Temperature of the vessels and oxygen consumption and pH are carefully controlled to optimize enzyme production.

Recovering the enzymes, referred to as harvesting, is done in a number of steps. First, the solids, referred to as biomass, are removed by filtration or centrifugation. The enzyme remains in the solution, called the broth. The broth is evaporated to concentrate the enzymes. The enzymes can be further purified by ion exchange, before further processing into powders, liquids, or granules. The residual biomass is stabilized by lime treatment and used as fertilizer. Table 12.1 gives a sampling of the major sources of enzymes.

HALAL CONTROL POINTS IN ENZYME PRODUCTION

Enzymes are harvested or extracted from natural biological sources such as animal tissues, plant materials, or microbial sources. Enzymes can also be produced through a fermentation process. Requirements for each process are quite different and unique to each product. Therefore, HCPs have been determined keeping the uniqueness of the processes in mind, as given in Figure 12.1 and Figure 12.2.

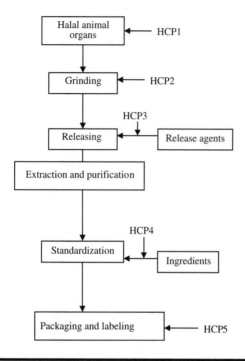

Figure 12.1   Conventional process for extracting enzymes of animal origins.

HALAL CONTROL POINTS FOR A CONVENTIONAL PROCESS FOR EXTRACTING ENZYMES FROM ANIMAL ORGANS

There are five HCPs: use of halal organs, clean equipment, acceptable release agents, approved standardization ingredients, and proper packing and labeling.

HCP1: Animal Organs

In commercial practice, enzymes are extracted from many animal organs of various species such as porcine, bovine, or ovine. Enzymes extracted from pig organs are not acceptable by any halal consumer or regulatory group, so these must not be used individually or in combination with other animal organs. Organs should be from halal-slaughtered animals to be universally acceptable. Such organs might have to be harvested from regions where Muslims perform animal slaughter. Some countries might accept enzymes from halal animals even though the animals are not slaughtered by Muslims.

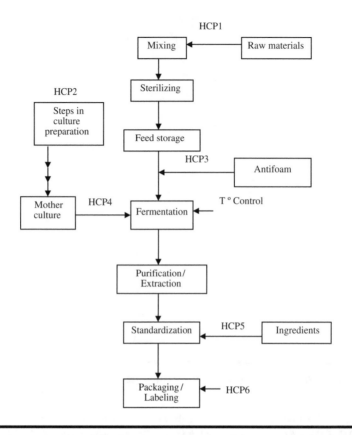

Figure 12.2 Enzyme production in a conventional fermentation process.

HCP2: Preparation of Tissues for Extraction

Because a majority of the enzymes of animal origin are from non-halal sources, use of common equipment presents a hazard. Before using the equipment for extracting halal enzymes, it must be thoroughly cleaned to avoid contamination from previous runs.

HCP3

Enzymes might not be present in the tissues in soluble form and therefore need to be released or made soluble to increase the yield. Chemicals used for this purpose must be suitable for halal production.

HCP4: Ingredients for Standardization

Besides salt and water, several other ingredients can be used to adjust the enzyme's strength. Preservatives and emulsifiers are also used to enhance or increase shelf life. All standardizing ingredients used should be suitable for halal production.

HCP5

Finally, labels should be marked properly, including halal markings to identify the product correctly. If the enzymes are from animals slaughtered by Muslims (dhabh procedure), they should be identified as such rather than labeling them simply bovine, etc. Packaging materials and labels themselves should conform to the guidelines discussed in Chapter 15.

HALAL CONTROL POINTS FOR ENZYME PRODUCTION IN A CONVENTIONAL FERMENTATION PROCESS

There are six critical control points in this process: raw materials, origin of cultures, acceptable processing aids, growth media, approved standardization ingredients, and packaging and labeling. Many of the process steps are similar to those shown in Figure 12.1, two major differences being the origin of cultures and growth media.

HCP1: Raw Materials for Growth Media

Growth media was not a major issue until seven employees of P.T. Ajinomoto were arrested in Indonesia for violating the country's law for manufacturing halal foods. In growth media, they used soy peptone that was manufactured with porcine enzymes (Roderick, 2001).

The first control point in the production of enzymes by fermentation is the use of acceptable media. Any raw materials even of vegetable origin should not have been modified or processed by using porcine or other non-halal enzymes or materials.

HCP2: Origin of Cultures

Microbial cultures, yeast, algae, or bacteria can be indigenous or be genetically engineered. All indigenous sources of culture are acceptable in commercial practice. However, if the bacteria or other microorganisms have been modified through biotechnology, the source of genetic material becomes important.

Genetic material from halal species of animals and all plant sources are generally acceptable. Food safety is a major concern but it is normally the responsibility of the government food regulatory agencies such as the FDA, USDA, and Environmental Protection Agency (EPA) in the U.S. and not of the religious regulatory agencies. As a general rule, all genetic materials from haram animals should be avoided.

HCP3

Processing aids such as antifoam should be free of prohibited materials, especially derivatives from pork fat.

HCP4

All materials used as growth media and in the preparation of the mother culture should be halal.

HCP5: Standardization Ingredients

Preservatives, emulsifiers, and other standardizing materials must be from acceptable sources. Alcohol may sometimes be used as a preservative to protect the enzyme activity. Alcohol is generally acceptable if below 0.5% by volume in the final enzyme preparation.

HCP6

Enzymes should be packed in acceptable containers and labeled properly with halal markings. For suitable packaging materials, see Chapter 15.

TO LABEL OR NOT TO LABEL ENZYMES

Enzymes are usually used as processing aids and functional catalysts. They are usually activated in the final product and are not included on the labels. For example, in fruit juice processing, enzymes are deactivated during pasteurization and might not be detected in the finished product, hence, not labeled. In cheese and bakery products, however, enzymes might remain active in the final product and are listed on the finished product label (Mannie, 2000).

Muslim consumers as well as halal authorities in many countries are concerned about the presence and source of enzymes. It is better to list not only the word enzymes but also their sources, even when the products are marked halal.

REFERENCES

Ashie, I.N.A. 2003. Bioprocess engineering of enzymes, *Food Technol.*, 57(1), 44-51.
Chaudry, M.M. 2002. Enzymes catalysts for life, *Halal Consum.*, No. 4, 5-7.
Mannie, E. 2000. Active enzymes, *Prep. Foods,* 169(10), 63-66, 68.
Mathewson, P.R. 1998. Major biological sources of enzymes (Appendix C), in *Enzymes*, Eagan Press, St. Paul, MN, pp. 93-95.

Olsen, H.S. 2000a. The nature of enzymes, in *Enzymes at Work*, Novozymes A/S Bagsvaerd, Denmark, p. 9.

Olsen, H.S. 2000b. Enzyme applications in the food industry, in *Enzymes at Work*, Novozymes A/S Bagsvaerd, Denmark, p. 25.

Roderick, D. 2001. Hold the pork, please, *Time*, 157(3), January 22.

13

ALCOHOL IN HALAL FOOD PRODUCTION

The word alcohol herein refers to ethanol, or ethyl alcohol, which is the main ingredient in what the Quran refers to as khamr, or alcoholic drinks. Alcohol is a chemical quite common in nature and has many uses and applications. In ancient times, it was mainly consumed as alcoholic drinks. Alcohol was made by fermentation from fruits such as grapes and dates, and is now also made from grains such as rye, wheat, barley, and corn. Potatoes and whey are also used to make alcohol.

The major uses of alcohol today are for alcoholic beverages and as a solvent in the food, cosmetics, and pharmaceutical industries. Alcoholic beverages legally can contain between 0.5 and 80% ethyl alcohol by volume. Pure industrial alcohol can be 95% alcohol.

MAJOR CATEGORIES OF ALCOHOLIC BEVERAGES

There are three major classes of alcoholic beverages:

- Fermented beverages — made from agricultural products, including grains and fruits, contain 3 to 16% alcohol.
- Distilled or spirit beverages — made by distillation of fermented beverages. Distillation increases alcohol content of these products up to 80%.
- Compound or fortified beverages — made by combining fermented or spirit beverages with flavoring substances. The alcohol content of these products can also be as high as 80%.

Alcoholic beverages can be consumed directly or added to foods, either as an ingredient during formulation or during cooking. When alcohol is an added ingredient, the ingredient label of the food product must list the specific alcoholic beverage that has been added and the final amount of alcohol if it is greater than 0.5%. Examples of this would be liqueur-flavored chocolates, cakes, and meals containing wines, such as beef stroganoff in wine sauce.

If a food contains more than 7% alcohol by volume, that food comes under the jurisdiction of the Bureau of Alcohol, Tobacco, and Firearms (BATF). FDA supervises any food that contains less than 7% alcohol. According to FDA, if alcohol is part of the food composition or formula, than alcohol must be included on the label as an ingredient. However, if alcohol is part of one of the ingredients, such as a flavor, it does not have to be listed separately on the ingredient label (Riaz, 1997).

Foods are cooked in alcohol to enhance the flavor or to impart a distinctive flavor. Wine is the most common form of alcohol used in cooking. Although it may seem that all of the added alcohol evaporates or burns off during cooking, it does not. Rena Cultrufelli of the USDA prepared a table listing the amount of retained alcohol in foods cooked in alcohol. The retained alcohol varies depending on the cooking method. The following are some of the retained alcohol contents of foods prepared by different cooking methods (Larsen, 1995):

- Added to boiling liquid and removed from the heat: 85%
- Cooked over a flame: 75%
- Added without heat and stored overnight: 70%
- Baked for 25 min without stirring: 45%
- Stirred into a mixture and baked or simmered for 15 min: 40%
- Stirred into a mixture and baked or simmered for 30 min: 35%
- Stirred into a mixture and baked or simmered for 1 h: 25%
- Stirred into a mixture and baked or simmered for 2 h: 10%
- Stirred into a mixture and baked or simmered for $2\frac{1}{2}$ h: 5%

Two of the major uses of pure alcohol are as a solvent and raw material. As a solvent, it is used to extract flavoring chemicals from plant materials such as vanilla beans. Dilute ethyl alcohol is almost universally used for the extraction of vanilla beans. After the extraction, vanilla flavor, called natural vanilla flavoring, is standardized with alcohol. By the FDA's standard of identity, natural vanilla flavoring must contain at least 35% alcohol by volume, otherwise it cannot be called natural vanilla flavoring (FDA, 2000).

As a raw material, one of the uses of alcohol is to convert it to acetic acid to make vinegar. Vinegar is then used in salad dressings, mayonnaise, and other applications. Whereas the use of alcohol in alcoholic drinks is haram, converting it to acetic acid (vinegar) makes it halal. This concept of conversion called chemical change or istihala is discussed in Chapter 16.

One important function of alcohol is to facilitate the mixing of oil-based ingredients into water-based products or water-based ingredients into oil-based products. This is an important use in the production of flavors. Most flavors are oils. For example, orange flavor is oil derived from orange skins. Orange flavor does not dissolve in water but dissolves in alcohol. The mixture of alcohol and orange flavor then dissolves in water. So to produce an orange-flavored carbonated drink, alcohol is used to make sure the orange flavor is fully mixed and dissolved in the carbonated water and remains dissolved over the expected shelf life of the product (Othman and Riaz, 2000).

Alcohol is also used in pharmaceuticals, cosmetics, and topical products. Alcohol is frequently present in cough syrups and mouthwash, though these days one can find some alcohol-free products. In perfumes, the use of SD alcohol is common. SD alcohol is ethanol that has been denatured. Denaturing involves adding substances to alcohol to make it undesirable for consumption. The denaturing substances are very difficult to remove from the mixture, so denatured alcohol cannot be used in food or drink, but the product remains prohibited for Muslims.

WHY ALCOHOL IS PROHIBITED FOR MUSLIMS

When the revelation of the Quran began, drinking alcoholic beverages was not uncommon. Many of the companions of the Prophet also drank prior to the prohibition in the Quran. The prohibition came in three stages. The first revelation is presented in the Quran [Arabic text and English rendering by Pickthall (1994)] as thus:

> *They question thee about strong drink and games of chance. Say: In both is great sin, and some utility for men; but the sin of them is greater than their usefulness. And they ask thee what they ought to spend. Say: That which is superfluous. Thus Allah maketh plain to you (His) revelations, that haply ye may reflect.*

> *Chapter II, Verse 219*

In this verse, Muslims were informed that drinking khamr is a great sin. It says that alcohol might have some benefit, although it imparts greater sin than benefit. The next revelation on the subject of alcohol was as follows:

> *Oh ye who believe! Draw not near unto prayer when ye are drunken, till ye know that which ye utter, nor when ye are polluted save when journeying upon the road, till ye have bathed. And if ye be ill, or on a journey, or one of you cometh from the closet, or ye have touched women, and ye find not water, then go to high clean soil and rub your faces and your hands (therewith). Lo! Allah is Benign, Forgiving.*

Chapter IV, Verse 43

In this revelation, Muslims are told not to offer their prayers while they are under the influence of alcohol, which will prevent them from understanding what they recite or listen to. Because prayer is prescribed five times a day, at specific times, one may conclude that it is nearly impossible to drink alcohol if one is to perform the required prayers on time in a sober condition. The verse still did not clearly say that alcohol was prohibited, so some companions continued to drink. Finally, a clear command was issued as follows, prohibiting intoxicants and several other actions:

> *O ye who believe! Strong drink and games of chance and idols and divining arrows are only an infamy of Satan's handiwork. Leave it aside in order that ye may succeed.*

Chapter V, Verse 90

> *Satan seeketh only to cast among you enmity and hatred by means of strong drink and games of chance, and to turn you from remembrance of Allah and from (His) worship. Will ye then have done?*

Chapter V, Verse 91

> *Obey Allah and obey the messenger, and beware! But if ye turn away, then know that the duty of Our messenger is only plain conveyance (of the message).*

Chapter V, Verse 92

This was the final word on alcohol as khamr — it must be avoided. There is no ambiguity: consumption of alcoholic drinks in any form is totally prohibited.

Although alcoholic drinks are prohibited for consumption, alcohol derived from grapes, dates, and raisins is also considered *najis* or unclean. Use of alcohol from these sources is not allowed even for nonfood uses, such as in cosmetics, perfumes, and personal care products. So alcohol is mainly haram as a food or unclean for other uses if it is from traditional sources of grapes and dates. Alcohol if made from halal sources can be converted into halal products. In contrast, pork is from pig, a haram source, and all products and ingredients derived from pork remain haram.

Alcoholic beverages of any type are prohibited for Muslims. The use of alcoholic beverages in preparing or producing food items or drinks is also prohibited. Hence, eating or drinking products made with alcoholic beverages, such as spiked punch, or cakes containing brandy, are not permitted. Grain alcohol or synthetic alcohol may be used in the production of food ingredients as long as it is evaporated to a final level of more than 0.5% in food ingredients and 0.1% in consumer products. These guidelines are practiced by some of the halal-certification organizations, whereas others follow somewhat stricter guidelines.

The following points serve as guidelines for the use of alcohol in halal food production:

- Natural products containing small amounts of intrinsic alcohol do not present a halal issue.
- Alcohol contained in a natural product may be concentrated into its essence, thereby concentrating the amount of alcohol. Most halal-certifying bodies accept small amounts of such inherent alcohol, generally less than 0.1% and sometimes up to 0.5%.
- Use of alcohol in any concentration in an industrial process is acceptable due to technical reasons where other viable alternatives are not available. The final alcohol content in the product of such industrial application must be reduced to less than 0.5% by evaporation or conversion to acetic acid. This means flavors that will be used in food production must not contain more than 0.5% alcohol to qualify as halal. Some countries, however, do accept amounts higher than 0.5%, whereas others have an even lower cut off.
- Addition of any amount of fermented alcoholic drinks such as beer, wine, or liquor to any food product or drink renders the product haram. However, if the essence is extracted from these products and alcohol is reduced to negligible amount, most halal-certifying

agencies and importing countries accept the use of such essences in food products. Consultation with proper authorities or end users can clarify this issue.

■ Consumer products with added ingredients that contain alcohol must have less than 0.1% alcohol, including both added and any natural alcohol, to qualify as halal. At this level, one cannot taste the alcohol, smell the alcohol, or see the alcohol, a criterion generally applied for the impurities. This reasoning has been established by the Islamic Food and Nutrition Council of America. Other groups may accept more lenient or stricter guidelines than these. The food industry should consult its customer companies or halal-approval agencies for their exact standard.

REFERENCES

FDA. 2000. 21 CFR 169.3(c).

Larsen, J. 1995. Ask the dietitian, Hopkins Technology, LLC Hopkins, MN, http://www.dietitian.com/alcohol.html.

Othman, R. and Riaz, M.N. 2000. Alcohol: a drink/a chemical, *Halal Consum.*, Fall Issue, 1, 17-21.

Pickthall, M.M. 1994. Arabic text and English rendering of *The Glorious Quran*, Library of Islam, Kazi Publications, Chicago, IL.

Riaz, M.N. 1997. Alcohol: the myths and realties, in *A Handbook of Halaal and Haraam Products*, Vol. 2., Uddin, Z., Ed., Publication Center for American Muslim Research and Information, Richmond Hill, NY.

14

FOOD INGREDIENTS IN HALAL FOOD PRODUCTION

This chapter discusses various types of single and compound ingredients, including spices, seasonings, condiments, sauces, dressings, batters, breadings, as well as curing agents, coatings, and flavorings. Thousands of ingredients are used in food and kindred industries. Most are generally recognized as safe (GRAS) or food additives, whereas a few are covered by other legal categories such as "prior-sanctioned" chemicals, that is, used before 1958, when the current law came into being. Some of the critical ingredients and processes from the halal perspective are listed and explained.

BACON BITS

Natural bacon bits are manufactured from real bacon made from pork. Artificial bacon bits are made with plant proteins, specifically soy protein, and colored and flavored with other ingredients. Although natural bacon bits are not acceptable for halal products, one may use artificial bacon bits as long as colors and flavors and other incidental ingredients, as well as the production equipment, are halal approved.

AMINO ACIDS

Many amino acids are used for various technical functions. They are either synthetic (made with starting materials of plant origin) or extracted from natural proteins. One of the most common and questionable amino acids is L-cysteine, which is used in pizza crust, doughnuts, and batters. Natural cysteine is a derivative from human hair, animal hair, duck feathers, and similar other sources. These days, L-cysteine is also made by using sugar as a base. L-Cysteine from human or animal hair is not acceptable as halal. However, it is acceptable if made from duck feathers, especially if the ducks have been slaughtered in an Islamic manner. Synthetic L-cysteine is also called vegetable grade, which is acceptable as halal as long as all the production requirements are properly carried out.

CIVET OIL

Civet oil is oil extracted from the glands of a cat-like animal called a civet. Civet oil is not accepted as halal.

LIQUOR

Liquors are alcoholic drinks commonly used in flavors either for taste or aesthetic appeal for people who enjoy drinking liquor. Use of liquor or any alcoholic drinks in preparation of flavorings or batters is not acceptable. One of the possible flavors for fried products or fried batter products is beer batter. Actual beer is used in the production of batter-coated fries, onion rings, or other fried appetizers. Although alcohol is flashed out during the frying process, beer batter still remains unacceptable as halal. This concept is discussed in detail in Chapter 13.

LIQUOR AND WINE EXTRACTS

The flavor industry has now created extracts of different types of wines and liquors to use strictly as flavoring agents. Such extracts may be used in formulating halal products as long as residual alcohol in the extracts is very low, generally less than 0.5%. One must consult the halal-certifying agency and the end user on whether there is a problem with the source.

FUSEL OIL DERIVATIVES

Some ingredients such as amyl alcohol or isoamyl alcohol are made from by-products of the alcohol industry. These are doubtful to use because halal agencies in certain countries do not approve of the use of such ingredients.

ENCAPSULATION MATERIALS

Several ingredients such as gelatin, cellulose, shellac, and zein are used to encapsulate food ingredients as well as food products. Shellac, cellulose, and zein are generally accepted in halal products. Gelatin is acceptable only if it is from halal-slaughtered animals or fish.

PRODUCTION OF VINEGAR FROM ALCOHOL

Vinegar is more than 90% water. Rather than trucking vinegar to manufacturing plants where salad dressings are manufactured, the companies tend to make vinegar onsite from alcohol. The alcohol used for this purpose is generally grain alcohol or sometimes called synthetic alcohol. This practice is generally acceptable but should be monitored to maintain the residual alcohol minimal level generally not to exceed 0.5%.

MINOR INGREDIENTS

Several of the GRAS ingredients can be used in the manufacture of any of the products discussed in this chapter. They can be used as carriers, solvents, anticaking agents, antidusting agents, or for any other technical function. They generally do not have a function in the final product. All these minor ingredients must also be from halal sources as defined and accepted by the halal-certifying agency.

MANUFACTURING OF FOOD PRODUCTS

Manufacturing of food products has become a very specialized field. Some companies specialize in producing pure food chemicals and additives. Others specialize in producing food ingredients of plant and animal origin. Still others create unique complex mixtures of ingredients, called proprietary mixes or secret formulas, such as flavorings. A finished food product purchased by a consumer can contain dozens of different ingredients. Manufacturers of the product generally do not purchase or use all of those ingredients individually. They may purchase several blends or unit packs to make the final product. For example, to produce fried fish, a manufacturer will buy fish from one supplier, oil from another supplier, and batter and breading from still other suppliers. Batter and breading are composed of many different ingredients for much different functionality, and they might be specifically developed for each manufacturer.

SPICES AND SEASONING BLENDS

The food industry has become highly specialized, where food processors use unit bags containing all the minor ingredients for convenience, economy, and uniformity, as well as for quality control. Spices and seasonings are single botanical ingredients or a dry blend of many different ingredients.

Seasoning manufacturers may use any of the food ingredients available to them whether they are of vegetable origin or animal origin.

There are two considerations in making halal seasoning blends. The first is composition of the blends. All components should be halal suitable. Noncertified animal-based ingredients should not be used in halal blends. The second consideration is cross-contamination from the equipment. Halal blend should be manufactured on thoroughly cleaned equipment or in dedicated mixers. Minor ingredients such as an encapsulating agent, antidusting agent, and free-flow agent must also be halal suitable.

CONDIMENTS, DRESSINGS, AND SAUCES

These are generally pourable or spoonable liquid products. Besides salt, they may contain oil, salt, sugar, and a number of other ingredients such as spices, flavors, acidulants, and preservatives. Most of the sauces, dressings, and condiments contain only vegetable ingredients or dairy ingredients. However, some of them can contain bacon bits, gelatin, wine, or complex flavorings. For halal production, such non-halal ingredients must not be included in the formulations. The product should be made on clean equipment; for the company's convenience, it is often preferable to do so earlier in the week after thorough cleaning of the equipment over the weekend.

BATTERS, BREADINGS, AND BREADCRUMBS

Batters, breadings, and breadcrumb manufacturing has also evolved into a specialized manufacturing process. These products are manufactured for performance according to their use in different products, whether the products are going to be deep fried, pan fried, or baked. Besides specialized flours and starches, other ingredients can be used for specific reasons.

L-Cysteine is used to modify the texture of batter or breading coatings. L-Cysteine must be either the vegetarian-type made through synthesis or from duck feathers from birds slaughtered according to halal requirements. Other minor ingredients used in batters, breadings, and breadcrumbs must meet the halal suitability requirements. It is generally acceptable to use nonalcoholic extracts as flavorings, but it is not acceptable to use beer or any other alcoholic drinks. Other ingredients of concern for halal production are coatings and sprays, which are used to keep the product from sticking to utensils, belts, and equipment.

FLAVORINGS

Flavors and flavorings are some of the most complex ingredients used in the food industry. Under U.S. regulations (FDA, 2002), individual components of a flavor need not be declared to customers. The flavor industry is exempt from revealing such information (which is considered to be proprietary) as long as the component ingredients of a flavor are either on an FDA or Flavor and Extract Manufacturers Association (FEMA) list. A flavor can contain any number of ingredients from one as simple as salt to those as complex as extracts and reaction ingredients.

Two groups of ingredients are of special concern to formulators of halal products: (1) unique flavoring agents such as civet oil and (2) ingredients of alcoholic origin. As a general guideline, ingredients of animal origin should be avoided in the development of flavorings unless those ingredients are halal certified.

It is permissible to use alcohol for extracting flavors or dissolving them. However, the amount of alcohol should be reduced to less than 0.5% in the final flavoring product. Certain countries or customers require lower allowances or even absence of alcohol for products brought into their countries.

Some countries do not permit fusel oil derivatives. It is advisable for formulators to work with their client companies and certifying agencies to determine the exact requirements of a certain company or country.

CHEESE FLAVORS

Dairy ingredients should be derived from processes that use either microbial enzymes or halal-certified animal enzymes.

MEAT FLAVORS

Meat and poultry ingredients should be from animals slaughtered according to halal requirements.

SMOKE FLAVORS AND GRILL FLAVORS

Halal concerns include the use of animal oils as a base for smoke and grill flavors or the use of emulsifiers from animal sources.

COLORANTS

Colors have been used in foods since antiquity (Francis, 1999). Historically, colors have been used not only to make food look appealing but also as adulterants to hide defects. Colors can be synthetic such as FD&C-certified colors, where water-soluble ones are called dyes and oil-soluble ones are called lakes. Colors can also be natural or organic, such as fruit and vegetable extracts, riboflavin, corn, corn endosperm, shellac, octopus ink, squid ink, chlorophylls, carotenoids, and caramel. Inorganic colors are also used in the food industry, such as titanium dioxide, carbon blue, iron oxide, and silicone dioxide. Some of these colors are used and sold in pure form, whereas others are a mixture of colorants and standardizing ingredients. Some of the ingredients believed to be used in colors are gelatin, emulsifiers, or antidusting agents. Formulators should use only halal-suitable standardization ingredients.

CURING AGENTS

Curing agents are specialized blends of salt, nitrites, and some other ingredients such as sodium ascorbate, sodium erythorbate, citric acid, and propylene glycol. Curing agents or curing mixtures are used for making sausage products. They are generally halal suitable. However, one should make sure that they are made on clean equipment and questionable ingredients are not incorporated into them.

COATINGS

Animal-based ingredients must be avoided. Particularly upsetting to Muslim consumers is the use of any haram ingredients for the coating of fruits and vegetables sold in the fresh produce section. The Nutritional Labeling and Education Act of 1990 contains specific provisions that require the packer to identify the source of components in any such coating and put it on the outside carton. Supermarkets and other retailers are required by law to have an easily visible sign indicating which fruits and vegetables in their store might have been treated with such coatings. These coatings might contain ingredients such as wax, beef tallow, petroleum wax, gelatin, sugars, zein protein, or any other ingredient to create a protective or cosmetic coating on the food product. Although vegetable and mineral ingredients used for this purpose are halal, the coating formulators must avoid doubtful ingredients such as beef tallow and gelatin, or haram ingredients such as lard. Sugars, zein, starches, bees wax, petroleum fractions, and vegetable oils are some halal-suitable ingredients for food coatings.

DAIRY INGREDIENTS

Ingredients such as dry milk powder, which are heat-processed, are suitable for halal manufacturing. Ingredients such as whey powder, lactose, whey protein isolates, and concentrates produced with the use of enzymes are questionable if the source of enzymes is unknown. To make these ingredients halal, producers must use microbial enzymes or enzymes from halal-slaughtered animal sources. Most certifying agencies require that dairy ingredients made with enzymes be halal certified.

This discussion is about a select list of ingredients and mixtures. A more detailed listing appears in the appendices. Use of E-numbers is common in Europe and Asian countries. Appendix M lists food ingredients with their E-designation and halal, haram, or doubtful status, and Appendix N lists ingredients and their halal, haram, or doubtful status.

REFERENCES

FDA. 2002. 21 CFR 101.22.
Francis, F.J. 1999. *Colorants*, Eagle Press, St. Paul, MN, p. 1.

15

LABELING, PACKAGING,
AND COATINGS
FOR HALAL FOOD

This chapter deals with many aspects of food production that wrap around the food in different ways. These varied aspects might seem unrelated and out of place here, but these are peripheral issues that touch the food and Muslim consumers' psyche one way or the other. The following topics will be briefly dealt with in this chapter:

- Labeling products for Muslim markets
- Halal issues in labels and printing
- Packaging food in a halal environment
- Packaging materials and containers
- Waxes, coatings, and edible films

LABELING PRODUCTS FOR THE MUSLIM MARKET

Labels are for the benefit of the consumer and should be quite descriptive, clear, and meaningful. Usually the ingredient label does not list the origin of the ingredients. Hidden ingredients such as processing aids, anticaking agents, carriers, and incidental ingredients from various sources present another serious problem for Muslim consumers. For example, magnesium or calcium stearates are used in the manufacturing of candy and chewing gum without mention of the origin of the stearates (Uddin, 1994). Some European manufacturers use up to 5% vegetable or animal fat in their product and are still able to label it pure chocolate. In many cases, it is not feasible to list every major and minor ingredient on the label. Halal certification of the product and proper halal markings and logos can clarify the doubt for consumers.

If alcohol is a part of the food composition or formulation, then alcohol must be included on the label as an ingredient. If alcohol is a part of other ingredients, then it comes under incidental ingredients. Some incidental additives are present in food at insignificant levels and do not have any technical or functional effect in the food and are exempt from food labeling requirements (Riaz, 1997). These two issues in labeling that concern Muslims, the hidden ingredients and the questionable ingredients, can both be addressed through halal certification and proper labeling and markings.

With placement of the certified halal symbol on food items, Muslims no longer need to memorize a list of mysterious E-numbers in Europe and chemical jargon in the U.S. each time they go shopping. It must be noted, however, that marked items are not the only halal items present, but the only ones that have been checked, confirmed, and certified.

One of the functions of food labeling regulations is to ensure that consumers receive adequate information about the food products to make the right personal choices, whether these choices are economic, philosophic, or health related. Through proper labeling, consumers can make a comparison between competing products, can avoid foods they may be allergic to, or, in many cases, verify the ingredient status as halal, haram, or doubtful. For example, if the label reads that a product contains lard, it is haram. In the same way, if the label reads that a product contains gelatin, it is doubtful because most of the time the source of gelatin is not revealed, although sometimes the company might reveal the source of gelatin as a fish or bovine source. If the food product containing gelatin is certified as halal by a reputable organization, Muslim consumers can buy the product without hesitation.

The information on a food package label in the U.S. can be divided into three types. The first is the mandatory information required by the Fair Packaging and Labeling Act and the Nutrition Labeling and Education Act, as well as the Food, Drug, and Cosmetic Act, and others (Potter and Hotchkiss, 1995). The second type of information is optional or voluntary information, but often regulated if present. The third is information that is provided by the manufacturer to help the consumer use or understand the product. This last type contains such information as instructions for preparation and additional recipes along with religious and philosophical certifications.

The information appearing on a label is as follows:

■ Food name — all food must be labeled on the principle display panel (PDP) with the common or usual name.

- Net quantity of content — tells consumers how much food is in the package or can.
- Ingredients — listing of all ingredients in descending order by weight is required by the labeling law except the ones that are less than 2% of the product, which could be listed in any order. If there is no halal marking or symbol on the label, then halal consumers generally look for information about the ingredients. But this remains a very unreliable way to judge food products and never deals with the condition in the manufacturing facility. Some of the minor ingredients as discussed earlier might not be revealed on the label and the product might have been manufactured on a product line where non-halal products were also made, for example, making canned peas on the same line as canned pork and beans.
- Company name — company name and address must be present on the label for consumer inquiries. This information helps Muslim consumers who care to write to the company to find out the status of several doubtful ingredients in the absence of halal markings. Company name and address should be complete and up-to-date for consumers, so that they can reach the company for information about the halal status of the ingredients or the halal status of the products.
- Product date — may be shown as expiration date, date food was packaged, or other codes.
- Nutrition information — nutrition-related information regulated on the packaging label are nutrition facts, nutrient content claims, and health claim (copy of nutrition information panel plus explanation).
- Other information — voluntary information provided by food companies, such as trademarks or copyright symbols and religious symbols, to indicate that the product has qualified for halal and kosher certification. Figure 15.1 shows a sample label with halal markings and Figure 15.2 one that is dual halal and kosher certified.
- Specific terminology — sometimes used to clarify words for a certain reason. "Red wine vinegar" might be a positive gourmet statement in the Western countries, but it is not viewed favorably by Muslim consumers, although there is no wine or significant amount of alcohol left in this product. Some people might think that all derivatives of alcohol/wine are haram and might not purchase a product containing wine vinegar (red or white). It is better to label the ingredients as flavored vinegar in these countries. Ingredients such as lecithin, mono- and diglycerides, and glycerin can be from animal or vegetable sources. Generally, the source is not identified on labels. If these ingredients were purely vegetable

(a) Front panel

(b) Back panel

Figure 15.1   A product label with halal markings (front and back panels). (Courtesy of Happy and Healthy Products, Inc., Boca Raton, FL.)

driven, then identifying them as vegetable mono- and diglycerides, soy lecithin, and plant glycerin, for example, would increase the chances of Muslim and vegetarian consumers. However, the best option available to the company is to have the product certified as halal and vegetarian by a reputable organization. Additionally, if the target market is other than English-speaking customers, then it will be prudent to make the labels bilingual or even multilingual, which should include appropriate marking from the halal-certifying agencies.

LABELS AND PRINTING ON FOOD

Paper and plastic labels, glue used for pressure-sensitive labels, hot-melt glues, edible printing dyes used directly on food, edible inks, and other similar issues seem very trivial. These materials might contain ingredients that are not permissible in halal food. Do they create a violation of halal

Figure 15.2 A halal and kosher certified product label. (Courtesy of My Own Meals, Inc., Deerfield, IL.)

guidelines if they seep into the food in minute quantities? Perhaps some of the halal-certifying agencies in other importing countries feel that such seepage might violate the halal status of food products. Some of the halal agencies, especially in importing countries, are concerned about such cross-contamination.

PACKAGING THE FOOD IN A HALAL ENVIRONMENT

In North America, halal food products are generally made in facilities that also produce non-halal products. Most of the workers in the production areas are non-Muslim, who are not familiar with halal. Manufacturers can work with their halal consulting agency to devise the following standard operating procedures to accommodate halal production requirements:

- Keeping halal products in a separate room
- Scheduling production to avoid cross-contamination
- Not switching workers from non-halal to halal packing areas
- Properly marking areas to identify halal production
- Ensuring that workers do not bring food into production areas, wash their hands before entering the facility, etc.

PACKAGING MATERIALS AND CONTAINERS

Some packaging materials are questionable as regards their halal status. In many cases, stearates of animal or vegetable origin might be used in production of plastic bags and containers. Waxes and coatings applied to plastic, paper, and Styrofoam cups and plates might be from animal fat following the very hot annealing stage, which is hot enough to nullify any animal products found prior to that step. Metal cans and drums can be contaminated with animal fats. Formation, rolling, and cutting of steel sheets to make containers requires the use of oils to aid in their manufacturing. Such oils can also be animal derived (Cannon, 1990). Steel drums, which are often reused, can be used to carry foods containing pork or pork fat, which, despite rigorous cleaning practices, might remain in small amounts to contaminate halal products otherwise thought pure.

EDIBLE COATING AND EDIBLE FILMS

Although the use of edible films in food products seems new, their use in the food industry actually started many years ago. In England, during the 16th century, larding, that is, coating food products with fat, was used to prevent moisture loss in foods (Labuza and Contrereas-Medellin, 1981). Currently, edible films and coatings find use in a variety of applications, including casing for sausages, chocolate coating for nuts and fruits, and wax coating for fruits and vegetables.

As the food industry develops edible films and coatings, it will be important for it to understand how some of the choices will affect the halal status of food products. This is particularly true for products that traditionally do not have an ingredient label and cannot be easily certified, such as fruits and vegetables. The Nutritional Labeling and Education Act has a series of regulations dealing with potential coatings of fruits and vegetables, including a mandatory label on packing boxes and signage at the point of sale. The wording of the labels was designed to give consumers meaningful category information, for example, vegetable, mineral, petroleum, lac-resin (shellac), or animal-based ingredients. All these categories except animal-based ingredients are acceptable for coatings. In addition, many consumers do not expect to find "animal" products on fresh or frozen fruits and vegetables. Thus, the selection of materials appropriate for these products in particular is important (Regenstein and Chaudry, 2002).

REFERENCES

Cannon, C. 1990. Islamic market spells opportunity for processor, *Nat. Provision.*, August, 17.

Labuza, T. and Contrereas-Medellin, R. 1981. Prediction of moisture protection requirements for foods. *Cereal Food World*, 26, 335-343.

Potter, N.N. and Hotchkiss, J.H. 1995. Governmental regulation of food and nutrition labeling, in *Food Science*, 5th ed., Chapman & Hall, New York, pp. 567-569.

Regenstein, J.M. and Chaudry, M.M. 2002. Kosher and halal issues pertaining to edible films and coating, in *Protein Based Film and Coating*, Gennadios, A., Ed., CRC Press, Boca Raton, FL.

Riaz, M.N. 1997. Alcohol: the myths and realties, in *A Handbook of Halaal and Haraam Products*, Vol. 2, Uddin, Z., Ed., Publication Center for American Muslim Research and Information, Richmond Hill, NY.

Uddin, Z. 1994. *A Handbook of Halaal and Haraam Products*, Center for American Muslim Research and Information, Richmond Hill, NY.

16

BIOTECHNOLOGY AND GMO INGREDIENTS IN HALAL FOOD

We have discussed halal laws in detail throughout this book. God requires us to eat of halal food (Pickthall, 1994):

> *(Saying): Eat of the good things wherewith We have provided you, and transgress not in respect thereof lest My wrath come upon you; and he on whom My wrath cometh, he is lost indeed.*

> *Chapter XX, Verse 81*

All things are considered good except the ones specifically prohibited, which are very few in number. We will look at genetically modified organisms (GMOs) and biotechnology in the light of some of the basic principles from Al-Qaradawi (1984) as discussed in Chapter 2.

Everything is halal unless specifically prohibited. There is no specific mention of altered, modified, genetically engineered food and ingredients in the Quran or the traditions of Muhammad, because these scientific developments are very recent. However, genetically modified or engineered products from prohibited animals are prohibited. For example, because pork is prohibited, by extension, any products made from genetically altered pigs are prohibited too.

- God is the only one who has the power to legislate for humans.
- A scientist can explain a new development, and a religious scholar can only try to interpret whether the development violates any of the tenets of Islam. Permitting haram and prohibiting halal is similar

to shirk, meaning ascribing partners to God. It would be most serious if GMOs were clearly haram, and Muslim scholars interpreted them as halal. This certainly is not the case.

- Haram is usually associated with what is harmful and unhealthy.
- If it is determined beyond doubt that any of the foods or ingredients developed through genetic modifications are harmful and unhealthy, they will not be approved by the government and the Islamic scholars will immediately declare them haram.
- There is always a better replacement for something that is haram. We have better replacements for haram ingredients through biotechnology. Until the mid-1980s, porcine pepsin was used in some cheese manufacture. Since the introduction of GM chymosin, the use of pepsin as a replacement for calf rennet has practically vanished. This is a big plus for biotechnology in the area of halal foods.
- To proclaim something halal that is not halal is also haram. Again, if genetically modified (GM) food were clearly haram, scholars would have a huge issue with it. The items that are *not* halal are clearly mentioned in the Quran and the traditions.
- Good intentions do not make haram into halal. This applies to pigs and other haram animals, even if scientists try to make pig cleaner and disease-free or grow pig organs for food in the lab; such organs are still haram.
- Doubtful things should be avoided. This is perhaps the most significant guideline. Muslims are required to avoid doubtful things. There is a clear tradition of the Prophet about avoiding doubtful things. If Muslim consumers feel that GM foods are doubtful, they must avoid them. Presently, doubtful GMOs are the ones modified with the use of the genes from prohibited animals.

Biotechnology is an extension of plant and animal breeding and genetics, which have been practiced for decades, and, in some cases, for centuries. One example of animal breeding dates back to prehistoric times when a donkey and a mare were crossbred to produce a mule. The meat of a donkey is not accepted as halal food, and therefore neither is the meat of a mule. Plants have always been bred with closely related plants and animals with closely related animals. Recently, genes were identified and scientists also learned how to take a gene from one species and move it to a more distant species. Currently, genes from fish or insects or pigs can be introduced into plant species without affecting the appearance or taste, but making the plants better resistant to diseases or nutritionally better compared with the conventional products available. This modern technology was not available at the inception of Islam. Muslim scholars are striving to come to an acceptable decision on some of the issues facing us today. At the inception of Islam, almost fourteen centuries ago, Islamic dietary laws were the only regulations for the safety and wholesomeness of food products, because there were no government food safety regulations. Currently, food safety is the responsibility of the government agencies and organizations such as the United Nation's Food and Agriculture Organization and World Health Organization. Issues relating to the safety of GM foods are deferred to such agencies; here we are only concerned with the religious aspects of GMOs. The underlying principle for halal is that food has to be halalun tayyaban, meaning permissible and wholesome, or good. Two government agencies in Malaysia, the Institut Kefahaman Islam Malaysia (IKIM) the and Jabatan Kemajuan Islam Malaysia (JAKIM), concur that GM food is halal as long as it is from halal sources using the halal methods of production (Kurien, 2002).

Several additional points can be considered here, such as the concept of change. Is there any change taking place in a gene transfer from a prohibited animal to a permitted animal? Does the gene change the character of the recipient animal or plant enough to make it prohibited? If not, then istihala (change of state) has taken place. Most GM products and ingredients fall within this concept. Is the porcine gene then acceptable? This is still a controversial issue.

Next is safety. Even if safe, if Muslim consumers feel that introducing pig genes into plants violates their religious responsibility, then such food is considered doubtful. Pork can be made safe by growing trichinae-free herds, but it remains haram. Any ingredients derived from pork used in food processing also make the food haram or at least doubtful for many Muslim consumers. The consumer has the right to accept or reject the reasoning behind the change. However, the industry, government agencies, and scholars have an obligation to educate the consumers about such issues.

The next point to be made is religious prohibition vs. personal inhibition. People might not find a basis to call the food prohibited because that is the right of God alone. But they might still not want to eat something because they are not sure or it makes them feel uneasy. This will not make GM foods haram.

Does the condition of necessity overrule prohibitions in the case of GM foods? Hunger is still prevalent in the world. GM foods certainly offer tangible alternatives. Most Americans believe that biotechnology will benefit them or their families in the next five years, according to a survey conducted by the International Food Information Council. Consumers expect benefits such as improved health and nutrition; improved quality, taste, and variety of foods; reduced chemical and pesticide use on plants;

reduced cost of food; and improved crops and crop yields (Langen, 2002). However, some European countries do not share these views.

Islam teaches caution and moderation to Muslims in eating food. GM foods and GM ingredients may not be haram, but many Muslims may avoid them anyway because they do not feel comfortable consuming them. The introduction of animal genes into plants presents a considerable ethical challenge and difficulties to consumers, Muslims and non-Muslims alike.

Theoretically, a donor gene can be from any of biological source, such as plants, microorganisms, insects, fish, or other animals. How does the source of donor gene affect the acceptability of the resultant GM products?

- A plant-to-plant gene transfer to make conventional ingredients such as citric acid and monosodium glutamate is acceptable.

- A plant-to-plant gene transfer where genetic material is actually consumed, such as tomatoes, corn, and rice, does not pose any issue either. These examples provide clear benefits, delayed ripening in the case of Flavr Savr tomatoes, protection against insects (corn borer) in the case of Bt corn, and enhancement in the nutritional value of golden rice (Nelson, 2001a, b; Nelson and Bullock, 2001).

- Transfer of animal genes to bacteria to manufacture enzymes and other bioactive ingredients is also acceptable as long as the safety of such ingredients is established beyond doubt and the production process is halal. Many of the enzymes produced these days use this new technology.

The complex gene products category deals with staple food products that have been modified through genetic engineering, with genes being transferred across species boundaries or reoriented within the same organism. If the genetic modifications improve product flavor, color, texture, shelf life, composition, etc., but do not alter the way such products are metabolized by the human body, and they are otherwise safe to consume, there might not be any acceptance problems by Muslims. However, if such products contain genes from prohibited animals, such as pigs, they might not be readily accepted. Use of porcine somatotropin for muscle mass buildup in beef cattle falls in this category (Chaudry and Regenstein, 1994).

GM rennet, called chymosin, is so widely used in cheese making that it has taken over the traditional sources of rennet. According to David Berrington of Chr. Hansen, Inc., about 80% of the cheese produced in the U.S. and U.K., and 40% of the world's cheese, is now made with genetically engineered chymosin (Avery, 2001). This process has been accepted as halal as long as the production is halal.

The use of genes from haram animals into halal animals or plants is going to be a very sticky issue. It will be difficult to convince Muslim consumers about the benefits of these GMOs. It is better for the industry to avoid such products. Other biotechnology-related issues such as cloning animals for food use and designing new species of animals are going to be equally challenging. There is no information available yet about how Muslim opinions on this issue will develop. It will be prudent for GM products with foreign genetic material present to be considered for review and evaluation by religious scholars.

A purely synthetic gene made through recombinant technology which is similar to a porcine lipase may be acceptable to produce halal GM products, because there is really no porcine material in the gene.

REFERENCES

Al-Qaradawi, Y. 1984. *The Lawful and Prohibited in Islam,* The Holy Quran Publishing House, Beirut, Lebanon.

Avery, D.T. 2001. Genetically modified organisms can help save the planet, in *Genetically Modified Organisms in Agriculture*, Academic Press, London, p. 211.

Chaudry, M.M. and Regenstein, J.M. 1994. Implication of biotechnology and genetic engineering for kosher and halal foods, *Trends Food Sci. Technol.*, 5, 165-168.

Kurien, D. 2002. Malaysia: studying GM foods' acceptability of Islam, August 9, Dow Jones Online News via News Edge Corporation, Kuala Lumpur, Malaysia.

Langen, S. 2002. IFIC conduct consumer food biotech survey, *Food Technol.*, 56(1), 12.

Nelson, G.C. 2001a. Introduction, in *Genetically Modified Organisms in Agriculture*, Academic Press, London, p. 3.

Nelson, G.C. 2001b. Traits and techniques of GMOs, in *Genetically Modified Organisms in Agriculture*, Academic Press, London, p. 9.

Nelson, G.C. and Bullock, D. 2001. The economics of technology adoption, in *Genetically Modified Organisms in Agriculture*, Academic Press, London, p. 17.

Pickthall, M.M. 1994. Arabic text and English rendering of *The Glorious Quran*, Library of Islam, Kazi Publications, Chicago, IL.

17

ANIMAL FEED
AND HALAL FOOD

There has been controversy recently surrounding the issue of using animal by-products and extracts in animal feed, mainly due to the crisis in the 1990s over the mad cow disease, also known as bovine spongiform encephalopathy (BSE). The status of feed formulated with animal ingredients needs to be considered from the halal perspective, that is, whether such feed nullifies the halal status of animals fed such feed. This chapter aims to clarify this issue and provide information to the industry.

The following discussion about the issue of feed mill inspection only points out deficiencies in implementation of the regulation. This information is meant for the food industry to understand that Muslim consumers are looking for products that are clean and wholesome in their perception. The animal feed issue is one such issue.

To help prevent the establishment and amplification of BSE through feed in the U.S., the FDA implemented a final rule that prohibits the use of most mammalian protein in feeds for ruminant animals. This rule, Title 21 Part 589.2000 of the Code of Federal Regulations, became effective on August 4, 1997. What does the above ruling mean by "most mammalian protein"? It seems that a certain amount of pork by-products and other mammalian by-products might still be used to formulate ruminant feed, as long as it meets the legal interpretation of the word *most* in the above ruling. Although active monitoring by the USDA has found no cases of BSE in U.S. cattle, compliance with the rule by the feed industry has indicated compliance problems. As part of the enforcement plan, an initial inspection assignment was issued to all FDA district offices in 1998 to conduct inspections of 100% of all renderers and known feed mills to determine compliance. These two segments of the feed industry are subject

Table 17.1 FDA Summary of Inspections

Category	Total number	Reported	Failed[a]	Status
Renderers	264	241	25	USDA licensed
Feed mills	1240 est.	1176	76	FDA licensed
Feed mills	6000–8000	4783	421	Not licensed
Other firms	Unknown	4094	110	Unknown

[a] Out of compliance.

to inspection and compliance under the BSE ruling. Besides renderers and feed mills, some of the other feed handlers, such as ruminant feeders, on-farm mixers, protein blenders, and distributors, were also included in the FDA inspections. The approximate number of all feed-handling firms in the U.S. along with data about the inspections is given in Table 17.1 (FDA, 2003).

Renderers are the first to handle rendered protein and send materials to feed mills and ruminant feeders. All 264 rendering firms were inspected, and results of 241 are available. All 1240 FDA-licensed feed mills have been inspected and results reported for 1176. Of the 6000 to 8000 feed mills that are not licensed, 4783 have been inspected or reported. Among the other firms handling ruminant feed, 4094 have been inspected or reported.

Various segments of the feed industry had different levels of compliance with the feed ban regulation, according to the FDA. As of June 12, 2001 (the latest date for which the data is available), 76% of the renderers inspected or reported (183 out of 264) were handling materials prohibited for use in ruminant feed (FDA, 2003). However, of those, only 25 (14%) had a problem, which indicated that they might not be properly segregating materials, either not labeling or comingling them according to the report.

Among all classes of firms handling ruminant feed, 27% of the firms inspected or reported (2653 out of 9867) were handling materials prohibited for use in ruminant feed (FDA, 2003). However, of those handling prohibited materials, 653 (25%) had a problem, which indicated that they might not be properly segregating materials, either not labeling or comingling them.

A number of observations and conclusions can be made from the data reported by the FDA to indicate that the possible inclusion of materials of concern to Muslims might still be occurring. Obviously, over time, this should become less of an issue for ruminants, but this does not cover other animals, particularly poultry.

- The ruling is applicable to ruminants (cattle) only and not to poultry.
- The ruling prohibits most, *not all*, mammalian by-products. (The report does not mention the specifics of by-products.)
- In three years, 100% of the feed handlers have not been inspected or reported.
- The majority of feed handlers handle prohibited materials.

Some of the firms failed inspections due to more than one reason.

It seems that some of the cattle farmers and almost all poultry farmers still have the opportunity to feed the animals "protein supplements" made from rendered animal parts, including swine. Although data about exact use of by-products is not available, let us suppose that the amount of by-products being fed to the ruminants has decreased as a result of the FDA ban. Hence, it is conceivable that slaughterhouse by-products from cattle and pigs are more available, which may be exported to other countries, including those with significant Muslim population. So the problem of by-products in feed is a universal one, not just a U.S. issue.

Current halal standards in the U.S. and other countries require that the animal be slaughtered according to Islamic guidelines but they do not control or emphasize farming conditions (Hussain, 2002). Misunderstanding on this subject by non-Muslims is understandable when Islamic scholars differ about what is "unclean" animal feed. Some scholars feel that haram animal parts fed to halal animals make them unclean and unfit for dhabh slaughter, whereas others believe that an animal has to live in filth and eat filth regularly to meet the condition of unclean. The Arabic word used for this concept is jalalah.

Jalalah refers to a condition of an animal living in and around heaps of filth and manure, and eating major portions of its feed out of such heaps of filth for a good part of its life. Muhammad forbade the consumption of meat and milk of such animals (Khan, 1991). It is important to keep in mind that all animals eat some dirt and filth even when they graze in green pastures and roam free on ranges. Rendered feed containing animal organs does not seem to fit this description of filth. It would seem something of an exaggeration to apply the rule of jalalah to animal feed in America and other countries.

Many Muslims feel very strongly that the feed for halal animals, whether raised for meat, milk, or eggs, must primarily be of plant origin, which the animals have been used to eating for centuries. Other Muslim consumers object only to pork by-products but not to formulated feed containing animal products per se. Last year, Saudi Arabia banned products from Europe on the suspicion that the animals were given feed containing prohibited animal parts (Al-Zobaidy, 2002).

It seems that the FDA ruling that makes it unlawful to include some mammalian by-products in feed for the ruminants does not go far enough in addressing the concerns of Muslim consumers about eliminating slaughterhouse by-products from ruminant and poultry feed. It would be prudent for the government and the industry to address this issue and create better guidelines than the ones that exist currently and fully enforce the ones that already exist.

REFERENCES

Al-Zobaidy, O. 2002. Imports of EU poultry, soft drinks banned, *Arab News*, July 27, www.arabnews.com/print.asp?id=.

FDA. 2003. www.fda.gov/cvm, Ruminant feed (BSE) enforcement activities, Center for Veterinary Medicine, Office of Management and Communications, Rockville, MD.

Hussain, M. 2002. Demand halal, consume halal, *HalalPak*, Summer Issue, 6-7.

Khan, G.M. 1991. *Al-Dhabh: Slaying Animals for Food the Islamic Way*, Abdul-Qasim Bookstore, Jeddah, Saudi Arabia.

18

COMPARISON OF KOSHER, HALAL, AND VEGETARIAN

Why are we discussing kosher and vegetarian issues in a halal book? There are two reasons:

- Permitted food of the Jews is called kosher. Many Muslims and non-Muslims think that halal is similar to kosher. We want to highlight the similarities and differences between kosher and halal. Similarly, many halal consumers might think that because vegetarian products are from vegetable sources, they are halal. Here, we point out differences not only between kosher and halal but also between halal and vegetarian.
- Many people in the U.S. food industry are familiar with the word kosher and what is required to make the products kosher. By comparing halal with kosher, we can help food industry professionals understand each concept and better comply with halal requirements.

Before undertaking a comparative discussion, let us first review the permissibility for Muslims of meat of animals killed by the Ahlul Kitab (Jews and Christians).

MEAT OF ANIMALS KILLED BY THE AHLUL KITAB

There has been much discussion among Muslim consumers as well as Islamic scholars about the permissibility of consuming meat of animals killed by the Ahlul Kitab (people of the book), meaning Jews and Christians. This generally implies that the animal was killed by Ahlul Kitab, but the Islamic method of slaughter while invoking the name of God as required under the Islamic guidelines was not followed.

In the holy Quran [Arabic text and English rendering by Pickthall (1994)] this issue is presented only once, in the following words:

> *This day are (all) good things made lawful for you. The food of those who have received the Scripture is lawful for you, and your food is lawful for them.*

> *Chapter V, Verse 5*

This verse addresses the Muslims and seems to have been set in a social context where Muslims, Jews, and Christians had to interact with each other. It points to two sides of the issue, first, "the food of the people of the book is lawful for you," and, second, "your food is lawful for them."

As regards the first part of the ruling, Muslims are allowed to eat the food of the Jews and Christians as long as it does not violate the opening statement of the verse, "This day all good and wholesome things have been made lawful for you."

The majority of Islamic scholars are of the opinion that the food of the Ahlul Kitab must meet the criteria established for halal and wholesome food, including proper slaughtering of animals. They believe that the following verse from the Quran establishes a strict requirement for Muslims.

> *And eat not of that whereupon Allah's name hath not been mentioned, for lo! It is abomination...*

> *Chapter VI, Verse 121*

However, some Islamic scholars such as Al-Qaradawi (1984) are of the opinion that this verse does not apply to the food of Ahlul Kitab. They opine that meat of halal animals sold in Western countries is acceptable for Muslims. They contend that God's name may be pronounced at the time of eating rather than at the time of slaughtering of an animal. Regulatory agencies in halal-food-importing countries, halal certifiers, or individual Muslim consumers can accept or reject products based on this reasoning.

For Muslims who want to follow the requirements of Chapter VI, Verse 121, none of the food of the Ahlul Kitab meets the Islamic standard, except vegetable items and fish if not prepared with alcohol or contaminated with prohibited ingredients.

According to Jackson (2000), most kosher food processors believe that Muslims accept kosher as meeting halal standards and requirement. Religiously, Muslims do not accept kosher certification as a substitute for halal certification. Although some countries did make allowances in the past, this is quickly changing (Jackson, 2000).

KOSHER LAWS

Kosher dietary laws determine which foods are fit or proper for consumption by Jewish consumers who observe these laws. The laws are Biblical in origin, coming mainly from the original five books of the Holy Scriptures (the Torah). At the same time Moses received the Ten Commandments on Mount Sinai, Jewish tradition teaches that he also received the oral law, which was eventually written down many years later in the Talmud. This oral law is as much a part of Biblical law as the written text. Over the years, the meaning of the Biblical kosher laws have been interpreted and extended by rabbis to protect Jewish people from violating any of the fundamental laws and to address new issues and technologies. The system of Jewish law is referred to as halacha (Regenstein and Chaudry, 2001).

HALAL LAWS

Halal dietary laws determine which foods are lawful or permitted for Muslims. These laws are found in the Quran and the books of hadith (the traditions). Islamic law is referred to as Shari'ah and has been interpreted by Muslim scholars over the years. The basic principles of Islamic laws remain definite and unaltered. However, their interpretation and application might change according to time, place, and circumstances. Some of the issues Muslim scholars are dealing with include biotechnology, unconventional sources of ingredients, synthetic materials, and innovations in animal slaughter and meat processing.

Although many Muslims purchase kosher food in the U.S., these foods, as we will see later, do not always meet the needs of Muslim consumers. The most common areas of concern for Muslim consumers when considering purchasing kosher products are the use of various questionable gelatins in products produced by more lenient kosher supervisions and the use of alcohol in cooking food and as a carrier for flavors.

KOSHER DIETARY LAWS

Kosher dietary laws predominantly deal with three issues, all focused on the animal kingdom:

- Allowed animals
- Prohibition of blood
- Prohibition of mixing of milk and meat

Additionally, for the week of Passover (in late March or April) restrictions on chometz, the prohibited grains (wheat, rye, oats, barley, and spelt), and rabbinical extensions of this prohibition lead to a whole new set of regulations, focused in this case on the plant kingdom. In addition, separate laws deal with grape juice, wine, and alcohol derived from grape products; Jewish supervision of milk; Jewish cooking, cheese making, and baking; equipment kosherization; purchasing new equipment from non-Jews; and old and new flour (Regenstein and Chaudry, 2001).

Allowed Animals for Kosher

Ruminants with split hoofs that chew their cud, traditional domestic birds, and fish with fins and removable scales are generally permitted. Pigs, wild birds, sharks, dogfish, catfish, monkfish, and similar species are prohibited, as are all crustacean and molluscan shellfish. Almost all insects are prohibited such that carmine and cochineal (natural red pigments) are not used in kosher products by most rabbinical supervisors. With respect to poultry, traditional domestic birds such as chicken, turkey, squab, duck, and goose are kosher. Birds in the rattrie category (ostrich, emu, and rhea) are not kosher as the ostrich is specifically mentioned in the Bible. However, it is not clear as to whether the animal of the Bible is the same animal we know today as an ostrich. A set of criteria is sometimes referred to in trying to determine whether a bird is kosher. The kosher bird has a stomach (gizzard) lining that can be removed from the rest of the gizzard. It cannot be a bird of prey. Another issue deals with tradition, for example, newly discovered or developed birds might not be acceptable. Some rabbis do not accept wild turkey, whereas some do not accept the featherless chicken.

The only animals from the sea that are permitted are those with fins and scales. All fish with scales have fins, so the focus is on scales. These must be visible to the human eye and must be removable from the fish skin without tearing the skin. A few fish remain controversial, probably swordfish being the most discussed (Regenstein and Regenstein, 2000).

Most insects are not kosher. The exception includes a few types of grasshoppers, which are acceptable in the parts of the world where the tradition of eating them has not been lost. Edible insects are all in the grasshopper family identified as permitted in the Torah due to their unique movement mechanism. Again, only visible insects are of concern; an insect that spends its entire life cycle inside the food is not of concern. The recent development of exhaustive cleaning methods to prepare prepackaged salad vegetables eliminates a lot of the insects that are sometimes visible, rendering the product kosher and, therefore, usable in kosher foodservice establishments and in the kosher home, without requiring extensive special inspection procedures. Although companies in this arena go through a great deal of effort to produce an insect-free product, some kosher supervision agencies remain unconvinced and only certify those products (or particular lots) that meet their more stringent requirements (Regenstein and Regenstein, 1988). The prohibition of insects focuses on the whole animal. If one's intent is to make a dish where the food will be chopped up in a food processor, then one may skip the elaborate inspection of fruits and vegetables for insects and assume that the presence of insect parts does not render the food non-kosher. There are guidebooks describing which fruits and vegetables in particular countries need inspection; recommended methods for doing this inspection are included. Kosher consumers have appreciated the use of pesticides to keep products insect-free as well as the use of prepackaged vegetables that have been properly inspected. Modern IPM (integrated pest management) programs that increase the level of insect infestation in fruits and vegetables can cause problems for the kosher consumer. Examples of problems with insects that one might not think about include insects under the triangles on the stalks of asparagus, under the greens of strawberries, and thrips on cabbage leaves. Because of the difficulty of properly inspecting them, many orthodox consumers do not use brussel sprouts (Regenstein and Regenstein, 1988).

Prohibition of Blood

Ruminants and fowl must be slaughtered according to Jewish law by a specially trained religious slaughterman (shochet), using a special knife designed for the purpose (chalef). The knife must be extremely sharp and have a very straight blade that is at least twice the diameter of the neck of the animal to be slaughtered. The animal is not stunned prior to slaughter. If the slaughter is done in accordance with Jewish law and with good animal-handling practices, the animal will die without showing any signs of stress. With respect to kashrus supervision, slaughtering is the only time a blessing is said, and it is said before commencing slaughter. The slaughterman asks forgiveness for taking a life. The blessing is not said over each animal. The rules for slaughter are very strict and the shochet checks the chalef before and after the slaughter of each animal. If any problem occurs with the knife, the animal becomes treife (not kosher). The shochet also checks the cut on the animal's neck after each slaughter to make sure it was done correctly. Slaughtered animals are subsequently inspected for defects by rabbinically trained inspectors. If an animal is found to have a defect, the animal is deemed unacceptable and becomes treife. There is no trimming of defective portions as generally permitted under secular law. The general rule is that the defect would not lead to a situation where the animal could be expected to die within a year. Consumer desire for more stringent kosher meat inspection requirements in the U.S. has led to the development of a standard for kosher meat that meets a stricter inspection requirement, mainly with respect to the condition of the animal's lungs. As the major site of halachic defects, the lungs must always be inspected. Other organs are spot-checked or examined when a potential problem is observed. Meat that meets this stricter standard is referred to as glatt (smooth) kosher, referring to the fact that the animal's lungs do not have any adhesions (sirkas). The bodek (inspector of internal organs) is trained to look for lung adhesions in the animal both before and after its lungs are removed. To test a lung, the bodek first removes all sirkas and then blows up the lung by using normal human air pressure. The lung is then put into a water tank and the bodek looks for air bubbles. If the lung is still intact, it is kosher. In the U.S., a glatt kosher animal's lungs generally have fewer than two adhesions, which permit the task to be done carefully in the limited time available in large plants (Regenstein and Chaudry, 2001; Regenstein and Regenstein, 1979, 1988).

Meat and poultry must be further prepared by properly removing certain veins, arteries, prohibited fats, blood, and the sciatic nerve. In practical terms, this means that only the front quarter cuts of kosher red meat are used in the U.S. and most Western countries. Although it is very difficult and time consuming to remove an animal's sciatic nerve, necessity demanded that this deveining be done in parts of the world where the hindquarter was needed in the kosher food supply. In some animals such as deer, it is relatively easy to devein the hindquarter. However, if there is no tradition of eating any hindquarter meat within a community, some rabbis have rejected the deer hindquarters for their community. To further remove the prohibited blood, red meat and poultry must then be soaked and salted within 72 h of slaughter. The salted meat is then rinsed three times (Regenstein and Chaudry, 2001; Regenstein and Regenstein, 1988). Any ingredients or materials that might be derived from animal sources are generally prohibited because of the difficulty of obtaining them from kosher animals. This includes many products that might be used in foods and dietary supplement, such as emulsifiers, stabilizers, and surfactants, particularly those materials that are derived from fat. Very careful rabbinical supervision is necessary to ensure that no animal-derived ingredients are included. Almost all such materials are available in a kosher form derived from plant oils (Regenstein and Chaudry, 2001).

Prohibition of Mixing of Milk and Meat

"Thou shalt not seeth the kid in its mother's milk." This passage appears three times in the Torah and is therefore considered a very serious admonition. The meat side of the equation has been rabbinically extended to include poultry. The dairy side includes all milk derivatives.

To keep meat and milk separate in accordance with kosher law requires that processing and handling of all materials and products fall into one of three categories:

- Meat product
- Dairy product
- Pareve (parve, parev), or neutral product

The pareve category includes all products that are not classified as meat or dairy. All plant products are pareve along with eggs, fish, honey, and lac resin (shellac). These pareve foods can be used with either meat products or dairy products. However, if they are mixed with meat or dairy, they take on the identity of the product they are mixed with; for example, an egg in a cheese soufflé becomes dairy.

To ensure the complete separation of milk and meat, all equipment, utensils, pipes, steam, etc., must be of the properly designated category. If plant materials (e.g., fruit juices) are run through a dairy plant, they will be considered a dairy product religiously. Some kosher supervision agencies permit such a product to be listed as dairy equipment (DE) rather than dairy. The DE tells the consumer that it does not contain any intentionally added dairy ingredients, but that it was made on dairy equipment (see discussion on allergy). If a product with no meat ingredients is made in a meat plant (e.g., a vegetarian vegetable soup), it may be marked meat equipment (ME). Although one may need to wash the dishes before and after use, the DE food can be eaten on meat dishes and the ME food on dairy dishes. A significant wait is normally required to use a product with dairy ingredients after one has eaten meat [i.e., from 3 to 6 h depending on the customs (minhag) of the area the husband came from]. With the DE listing, the consumer can use the DE product immediately before or after a meat meal but not with a meat meal. Following dairy, the wait before eating meat is much less, usually from a rinse of the mouth with water to 1 h. Certain dairy foods require the full wait of 3 to 6 h; for example, when a hard cheese (defined as a cheese that has been aged for over 6 months or one that is particularly dry and hard, such as many of the Italian cheeses) is eaten, the wait is the same as that for meat to dairy. Thus, most companies producing cheese for the kosher market usually age their cheese for less than 6 months, although with proper package marking this is not a religious requirement. If one wants to make an ingredient or product truly pareve, the plant equipment must undergo a process of equipment kosherization (Regenstein and Chaudry, 2001).

Kosher: Special Foods

Rules governing grape products, yashon flour, Jewish milk, and other foods are beyond the scope of this chapter.

Passover Requirements

The Passover holiday comes in spring and requires observant Jews to avoid eating the usual products made from five prohibited grains: wheat, rye, oats, barley, and spelt (Hebrew: chometz). Those observing kosher laws can eat only the specially supervised unleavened bread from wheat (Hebrew: matzos) that is prepared especially for the holiday. Once again, some matzos (schmura matzos) are made to a stricter standard with rabbinical inspection beginning in the field. For other Passover matzo, supervision does not start until the wheat is about to be milled into flour. Matzo made from oats and spelt is now available for consumers with allergies.

Special care is taken to ensure that the matzo does not have any time or opportunity to rise. In some cases, this literally means that products are made in cycles of less than 18 min. This is likely to be the case for handmade schmura matzo. In continuous large-scale operations, the equipment is constantly vibrating so that there is no opportunity for the dough to rise (Regenstein and Chaudry, 2001; Regenstein and Regenstein, 1988).

Equipment Kosherization

There are three ways to make equipment kosher or to change its status back to pareve from dairy or meat. (Rabbis generally frown on going from meat to dairy or vice-versa. Most conversions are from dairy to pareve or from treife to one of the categories of kosher.) There are a range of process procedures to be considered, depending on the equipment's prior production history.

HALAL DIETARY LAWS

Halal dietary laws deal with the following five issues; all except one are in the animal kingdom:

- Prohibited animals
- Prohibition of blood
- Method of slaughtering and blessing
- Prohibition of carrion
- Prohibition of intoxicants

Islamic dietary laws are derived from the Quran, a revealed book; the hadith, the traditions of Muhammad; and through extrapolation of and deduction from the Quran and the hadith by Muslim jurists.

The Quran states:

> *Forbidden Unto you (for food) are carrion and blood and swine-flesh, and that which hath been dedicated unto any other than Allah, and the strangled, and the dead through beating, and the dead through falling from a height, and that which hath been killed by (the goring of) horns, and the devoured of wild beasts, saving that which ye make lawful (by the death-stroke), and that which hath been immolated unto idols. And (forbidden is it) that ye swear by the divining arrows. This is an abomination. This day are those who disbelieve in despair of (ever harming) your religion; so fear them not, fear Me! This day have I perfected your religion for you and completed my favor unto you, and have chosen for you as religion AL-ISLAM. Whoso is forced by hunger, not by will, to sin: (for him) Lo! Allah is Forgiving, Merciful.*

Chapter V, Verse 3

The Quran also states:

> *O ye who believe! Eat of the good things wherewith We have provided you, and render thanks to Allah, if it is (indeed) He whom you worship.*

Chapter II, Verse 172

Eleven generally accepted principles pertaining to halal (permitted) and haram (prohibited) in Islam provide guidance to Muslims in their customary practices:

- The basic principle is that all things created by Allah are permitted, with a few exceptions that are prohibited. Those exceptions include pork, blood, meat of animals that died of causes other than proper slaughtering, food that has been dedicated or immolated to someone other than God, alcohol, and intoxicants.
- To make lawful and unlawful is the right of Allah alone. No human being, no matter how pious or powerful, may take it into his or her own hands to change things.

■ Prohibiting what is permitted and permitting what is prohibited is similar to ascribing human partners to Allah. This is a sin of the highest degree that makes one fall out of the sphere of Islam.

■ The basic reasons for the prohibition of things are impurity and harmfulness. A Muslim is not supposed to question exactly why or how something is unclean or harmful in what Allah has prohibited. There might be obvious reasons and there might be obscure reasons. To a person of scientific mind, some of the obvious reasons can be as follows:

  ■ Carrion and dead animals are unfit for human consumption because the decay process leads to the formation of chemicals harmful to humans (Awan, 1988).

  ■ Blood that is drained from an animal contains harmful bacteria, products of metabolism, and toxins (Hussaini and Sakr, 1983).

  ■ Swine serves as a vector for pathogenic worms to enter the human body. Infections by *Trichinella spiralis* and *Traenia solium* are not uncommon (Awan, 1988).

  ■ The fatty acid composition of pork fat has been mentioned as incompatible with human fat and human biochemical systems (Sakr, 1991).

  ■ Intoxicants are considered harmful for the nervous system, affecting the senses and human judgment, leading to social and family problems, and might even lead to death (Al-Qaradawi, 1984; Awan, 1988).

    These reasons and other similar explanations may sound reasonable to a layperson but become more questionable under scientific scrutiny. If meat of dead animal were prohibited due to harmful chemicals in decaying meat, then dead fish would have been prohibited. If pork contains *Trichinae,* beef might contain *E. coli.* If pork fat is bad, so are trans fatty acids. The underlying principle, it seems, behind the prohibitions is not scientific reasons but the divine order "forbidden unto you are…."

■ What is permitted is sufficient and what is prohibited is then superfluous. Allah prohibited only things that are unnecessary or dispensable, while providing better alternatives. People can survive and live better without consuming unhealthful carrion, unhealthful pork, unhealthful blood, and the root of many vices — alcohol.

■ Whatever is conducive to the prohibited is in itself prohibited. If something is prohibited, anything leading to it is also prohibited.

■ Falsely representing unlawful as lawful is prohibited. It is unlawful to make flimsy excuses, to consume something that is prohibited, such as drinking alcohol for supposedly medical reasons.

- Good intentions do not make the unlawful acceptable. Whenever any permissible action of the believer is accompanied by a good intention, his action becomes an act of worship. In the case of haram, it remains haram, no matter how good the intention or how honorable the purpose. Islam does not endorse employing a haram means to achieve a praiseworthy end. Islam indeed insists that not only the goal be honorable, but also the means chosen to achieve it be lawful and proper. Islamic laws demand that right should be secured through just means only.
- Doubtful things should be avoided. There is a gray area between clearly lawful and clearly unlawful. This is the area of "what is doubtful." Islam considers it an act of piety for the Muslims to avoid doubtful things, for them to stay clear of unlawful. Prophet Muhammad said: "The halal is clear and the haram is clear. Between the two there are doubtful matters concerning which people do not know whether they are halal or haram. One who avoids them in order to safeguard his religion and his honor is safe, while if someone engages in a part of them, he may be doing something haram."
- Unlawful things are prohibited to everyone alike. Islamic laws are universally applicable to all races, creeds, and sexes. There is no favored treatment of a privileged class. Actually, in Islam, there are no privileged classes; hence, the question of preferential treatment does not arise. This principle applies not only among Muslims, but between Muslims and non-Muslims as well.
- Necessity dictates exceptions. The range of prohibited things in Islam is quite limited, but emphasis on observing these prohibitions is very strong. At the same time, Islam is not oblivious to the exigencies of life, to their magnitude, or to human weakness and capacity to face them. A Muslim is permitted, under the compulsion of necessity, to eat a prohibited food in quantities sufficient to remove the necessity and thereby survive (Regenstein and Chaudry, 2001, Riaz, 1999a; Chaudry, 1992).

Prohibited and Permitted Animals

Meats of pigs, boars, and swine are strictly prohibited, and so are meats of carnivorous animals such as lions, tigers, cheetahs, dogs, and cats, and birds of prey such as eagles, falcons, ospreys, kites, and vultures.

Meat of domesticated animals such as ruminants with split hoof, such as cattle, sheep, goat, and lamb, is allowed for food, and so are meats of camels and buffaloes. Also permitted are meats of birds that do not use their claws to hold down food, such as chickens, turkeys, ducks, geese, pigeons, doves, partridges, quails, sparrows, emus, and ostriches. Some of the animals and birds are permitted only under special circumstances or with certain conditions. The animals fed unclean or filthy feed such as feeds formulated with sewage (biosolids) or protein from dead animals must be quarantined and placed on clean feed for three to forty days (Awan, 1992).

Foods from the sea, namely fish and seafood, are the most controversial among various denominations of Muslims. Certain groups accept only fish with scales as halal, while others consider everything that lives in water all the time or some of the time, as halal. Consequently, prawns, lobsters, crabs, and clams are halal for most Muslims but may be detested (makrooh) by some and hence not consumed.

There is no clear status for insects established in Islam except that locust is specifically mentioned as halal. Among the by-products from insects, use of honey was very highly recommended by Muhammad. Other products such as royal jelly, wax, shellac, and carmine are acceptable to be used without restrictions by most Muslims; however, some might consider shellac and carmine makrooh or offensive to their psyche.

Eggs and milk from permitted animals are also permitted for Muslim consumption. Milk from cows, goats, sheep, camels, and buffaloes is halal. Unlike kosher, there is no restriction on mixing meat and milk (Regenstein and Chaudry, 2001; Chaudry, 1992).

Prohibition of Blood

According to Quranic verses, blood that pours forth from an animal when it is slaughtered is prohibited for consumption. It includes blood of permitted and nonpermitted animals alike. Liquid blood is generally not offered for sale or consumed even by non-Muslims, but products made with and from blood, such as blood plasma proteins, are available. There is general agreement among Muslim scholars that anything made from blood of any animal, including fish, is unacceptable. Products such as blood sausage and ingredients such as blood albumin are either haram or questionable at best, and should be avoided in product formulations (Riaz, 1996).

Proper Slaughtering of Permitted Animals

There are special requirements for slaughtering animals:

- Animal must be of a halal species.
- Animal must be slaughtered by an adult and sane Muslim.

- The name of Allah must be pronounced at the time of slaughter.
- Slaughter must be done by cutting the throat in a manner that induces rapid and complete bleeding, resulting in the quickest death. The generally accepted method is to cut at least three of the four passages in the neck, that is, carotids, jugulars, trachea, and esophagus.

The meat of animals thus slaughtered is called zabiha or dhabiha meat.

Islam places great emphasis on gentle and humane treatment of animals, especially before and during slaughter. Some of the conditions include giving the animal proper rest and water, avoiding conditions that create stress, not sharpening the knife in front of the animals, and using a very sharp knife to slit the throat. After the blood is allowed to drain completely from the animal and the animal has become lifeless, only then may dismemberment begin, such as cutting off horns, ears, or legs. Unlike kosher, postslaughter religious inspection, deveining, and soaking and salting of the carcass is not required for halal. Hence, halal meat is treated no different than commercial meat. Animal-derived food ingredients such as emulsifiers, tallow, and enzymes must be made from animals slaughtered by a Muslim to be halal.

Hunting of permitted wild animals, such as deer and elk, and birds, such as doves, pheasants, and quails, is permitted for the purpose of eating, but not merely for deriving pleasure out of killing an animal. Hunting by any means by tools such as guns, arrows, spears, or trapping is permitted. Trained dogs or birds of prey may also be used for catching or retrieving the hunt as long as the hunting animal does not eat any of the prey. The name of Allah may be pronounced at the time of ejecting the hunting tool rather than the actual catching of the hunt. The hunted animal has to be bled by slitting the throat as soon as it is caught. If the blessing is made at the time of pulling the trigger or the shooting of an arrow and the hunted animal dies before the hunter reaches it, it would still be halal as long as slaughter is performed and some blood comes out. Fish and seafood may be hunted or caught by any reasonable means available as long as it is done humanely.

The requirements of proper slaughtering and bleeding are applicable to land animals and birds. Fish and other creatures that live in water need not be ritually slaughtered. Similarly, there is no special method of killing locust.

The meat of a permitted animal that dies of natural causes or diseases, from being gored by other animals, by being strangled, by falling from a height, through beating, or by being killed by wild beasts is unlawful to be eaten unless one saves such animals by halal slaughtering before they actually become lifeless. Fish that dies of itself, if floating on water or if lying out of water, is still halal as long as it does not show any signs of decay or deterioration.

An animal must not be slaughtered after dedication to someone other than Allah and immolated to anybody other than Allah under any circumstances. This is a major sin (Regenstein and Chaudry, 2001; Chaudry and Regenstein, 2000; Chaudry, 1992).

Prohibition of Alcohol and Intoxicants

Consumption of alcoholic drinks and other intoxicants is prohibited according to the Quran as follows:

O ye who believe! Strong drink and games of chance and idols and divining arrows are only an infamy of Satan's handiwork. Leave it aside in order that ye may succeed.

Chapter V, Verse 90

Satan seeketh only to cast among you enmity and hatred by means of strong drink and games of chance, and to turn you from the remembrance of Allah and from (His) worship. Will ye then have done?

Chapter V, Verse 91

The Arabic term used for alcohol in the Quran is khamr, which means that which has been fermented, and implies not only to alcoholic beverages such as wine, beer, whiskey, or brandy but also to all things that intoxicate or affect one's thought process. Although there is no allowance for added alcohol in any beverage such as soft drinks, the small amount of alcohol contributed from food ingredients might be considered an impurity and hence ignored. Synthetic or grain alcohol can be used in food processing for extraction, precipitation, dissolving, and other reasons as long as the amount of alcohol remaining in the final product is very low, generally below 0.1%. However, each importing country has its own guidelines that must be understood by exporters and strictly adhered to (Regenstein and Chaudry, 2001; Riaz, 1997; Chaudry, 1992).

Halal Cooking, Food Processing, and Sanitation

There are no restrictions about cooking in Islam, as long as the kitchen is free from haram foods and ingredients. There is no need to keep two sets of utensils, one for meat and the other for dairy, as in kosher.

In food companies, haram materials should be kept segregated from halal materials. The equipment used for non-halal products has to be thoroughly cleansed by using proper techniques of acids, bases, detergents, and hot water. As a general rule, kosher clean-up procedures are adequate for halal too. If the equipment is used for haram products, it must be properly cleaned, sometimes by using an abrasive material, and then be blessed by a Muslim inspector by rinsing it with hot water seven times (Regenstein and Chaudry, 2001).

KOSHER AND HALAL

Gelatin

Important in many food products, gelatin is probably the most controversial of all modern kosher and halal ingredients. Gelatin can be derived from pork skin, beef bones, or beef skin. In recent years, some gelatins from fish skins have also entered the market. As a food ingredient, fish gelatin has many similarities to beef and pork gelatin, such as a similar range of bloom strengths and viscosities. However, depending on the species from which fish skins are obtained, its melting point can vary over a much wider range of melting points than beef or pork gelatin. This offers some unique opportunities to the food industry, especially for ice cream, yogurt, dessert gels, confections, and imitation margarine. Fish gelatins can be produced kosher and halal with proper supervision, and be acceptable to almost all the mainstream religious supervision organizations.

Most of the currently available gelatin — even if called kosher — is not acceptable to the mainstream U.S. kosher supervision organizations and to the Muslim community. Many gelatins are, in fact, totally unacceptable to halal consumers because they might be pork based.

A recent development has been the manufacture of kosher gelatin from the hides of kosher-slaughtered cattle. It has been available in limited supply at great expense, and this gelatin has been accepted by the mainstream and even some of the stricter kosher supervision agencies. The gelatin companies produce gelatins of different bloom strengths and both soft and hard capsules of various sizes. This is an important new development that should be of interest to neutraceutical and drug markets. Similarly, at least two major manufacturers are currently producing certified halal gelatin from cattle bones of animals that have been slaughtered by Muslims. Halal-certified hard and soft gelatin capsules are available at competitive prices. Hard, two-piece and soft, one-piece capsules made with different vegetable materials are also available, most of which are certified as halal, kosher, and vegetarian.

One finds a wide range of attitudes toward gelatin among the lenient kosher supervision agencies. The most liberal view holds that gelatin, being made from bones and skin, is not being made from a food (flesh). Further, the process used to make the product goes through a stage where the product is so unfit that it is not edible by humans or dogs, and as such becomes a new entity. Rabbis holding this view might accept pork gelatin. Most water gelatin desserts with a generic K follow this ruling.

Other rabbis permit gelatin only from beef bones and hides, and not pork. Still other rabbis only accept "Indian dry bones" as a source of beef gelatin. These bones, found astray in India, are aged and become degreased over time and are considered "dry as wood" by rabbis. Kosher religious laws exist for permitting these materials. Again, none of these products is accepted by the mainstream kosher or halal supervisions, and therefore not accepted by a significant part of the kosher and halal community (Regenstein and Chaudry, 2001; Riaz, 1999b).

Biotechnology

Rabbis and Islamic scholars currently accept products made by simple genetic engineering; for example, chymosin (rennin) was accepted by rabbis about a half a year before the FDA accepted it. Production conditions in fermenters must still be kosher or halal; that is, the ingredients, the fermenter, and any subsequent processing must use kosher or halal equipment and ingredients of the appropriate status. A product produced in a dairy medium, for example, extracted from cow's milk, is dairy. Mainstream rabbis may approve porcine lipase made through biotechnology when it becomes available, if all the other conditions are kosher. The Muslim community is still considering the issue of products with a porcine gene; although a final ruling has not been established, the leaning seems to be toward rejecting such materials. Religious leaders of both communities have not yet determined the status of more complex genetic manipulations (Regenstein and Chaudry, 2001; Riaz, 1999a; Chaudry and Regenstein, 1994).

VEGETARIANISM

Vegetarianism encompasses a variety of options and choices, based on life styles, philosophies, and religions. The preferences vary from eating nothing but the parts of plants that be picked without destroying the plant to eating everything except flesh (red meat). Types of vegetarians from lenient to the most strict include (The Vegetarian Society, 2002):

- Pesco vegetarians — eat fish, eggs, and dairy products, but avoid poultry and meat products.

- Lacto-ovo vegetarians — consume all types of vegetable products, eggs, and milk products, but avoids all forms of slaughtered flesh, including meat, poultry, and fish. People who do not eat eggs but eat dairy products are called lacto-vegetarians, whereas ovovegetarians consume eggs but not dairy products.
- Vegans — do not eat anything of animal origin. A vegan therefore avoids all meats, poultry, and any other animal products and their derivatives, such as gelatin; eggs; milk, cheese, yogurt, and other dairy products; and fish, shellfish, crustaceans, and other marine animal products. Vegans also try to avoid honey, royal jelly, and cochineal and other insect-derived products. In addition, vegans do not knowingly consume hidden animal ingredients (The Vegan Society, 2002).

Fruitarians and foodists follow a type of vegan diet of fruits, vegetables, seeds, and nuts that is minimally processed or cooked. Fruitarians believe that only plant foods that can be harvested without killing the plant should be eaten (The Vegan Society, 2002).

Vegetarianism is not a particular religion. Believers of many religious denominations including Muslims, Jews, Seventh Day Adventist, Christians, Latter Day Saints (Mormons), Hindus, Buddhists, and Jains might practice vegetarianism to some extent.

Mainstream vegetarianism is usually defined as lacto-ovo; however, veganism is becoming quite popular in the West. The comparison to be made here is limited to vegan and lacto-ovo vegetarianism rather than other types of vegetarianism.

Food Standards for Vegetarians

Animal Products

For both vegans and lacto-ovo vegetarians, all products of animal flesh including food ingredients from meat, poultry, and seafood must be avoided. Moreover, for vegans, products and by-products from live animals, such as milk and eggs, and products from insects, such as honey and royal jelly, must also be avoided. Lacto-ovo vegetarians who consume egg and dairy products avoid incidental ingredients of animal origin, including enzymes such as rennet. However, most of the enzymes for cheese manufacture at least in the U.S. are from microbial or GM origin, which has found acceptance among this group.

Ingredients not acceptable as vegetarian include Vitamin D from sheep wool, gelatin in juice processing, and anchovies in Worcestershire sauce (The Vegan Society, 2002).

Kitchen and Hygiene Standards

Dishes and utensils used for preparing and serving vegetarian products must be separate from nonvegetarian dishes or at a minimum must be thoroughly washed. It is recommended that cross-contamination from nonvegetarian food be avoided.

COMPARISON OF KOSHER, HALAL, AND VEGETARIAN

The viewpoints expressed here are those of Orthodox movement and not of the Conservative or Reform movements. Table 18.1 gives a full comparison of kosher, halal, and vegetarianism. There are certain similarities and many differences, especially between kosher and halal. In both religions, pork and pork products derived from pigs and swine are prohibited. Also, carnivorous animals and birds are not allowed in either religion. Because it is understood that animal products are against the philosophy of vegetarianism, there is no need to mention vegetarianism specifically in this comparison.

Of the permitted animals, ruminants and poultry have to be killed by a Jew to make them kosher and killed by a Muslim to make them halal. In kosher, all animals have to be hand-slaughtered without stunning. However, an animal may be stunned after slaying to facilitate bleeding in some cases, but it is not acceptable as glatt kosher. For halal, stunning by captive bold or electrical is permitted as long as the animal is alive at the time of slaying. For halal, many countries accept mechanical slaughter of poultry if supervised by a Muslim with the back-up person being a Muslim.

In kosher, a blessing is made before entering the kill floor and start of the slaughter operation. It is not proper to invoke the name of God out of context, and in a dirty place such as a slaughterhouse. However, for halal, an invocation of the name of God must be made at the time of slaying an animal. Only front quarters of ruminants are used for kosher meat. Meat is also soaked in water, and covered with salt for an hour to remove the blood. For halal, the entire animal is permitted to be consumed. There is no requirement for soaking or salting of meat for halal. Blood and blood by-products are not permitted under kosher as well as halal rules.

Among fish and seafood, only fish with fins and removable scales is kosher, whereas all fish and seafood are halal. However, some Muslim denominations do not accept fish without scales or seafood, or both, as halal.

Enzymes derived from microbial or biotech (GMO) sources are acceptable as kosher, halal, and vegetarian. Enzymes extracted from kosher-killed animals are accepted as kosher and enzymes extracted from halal-killed animals accepted as halal. Some liberal rabbis might accept enzymes from non-kosher-killed animals as well as porcine enzymes. Enzymes from non-halal-killed animals might be accepted by some groups and countries, but not all. Porcine enzymes are generally not accepted by Muslims (Riaz, 1999a).

Gelatin from kosher-killed animals is accepted by all Orthodox rabbis as kosher and gelatin from halal-slaughtered animals is acceptable as halal. Some countries also accept regular bovine gelatin under certain conditions. Gelatin from fish with scales only is kosher, but gelatin from any fish is halal.

For dairy ingredients and cheese, the reader is referred to the discussion on enzymes. Additionally, for cheese to be kosher, a Jew must add the cultures to the milk for mainstream kosher. There is no such restriction for halal, in which any person may add the cultures to the milk.

Alcohol, especially alcoholic drinks, is not allowed for Muslims, but Jews consider alcohol as kosher; however, some restrictions exist about the source of alcoholic drinks in kosher. All alcoholic products are vegetarian. It is also important for kosher *not* to combine meat and dairy products, but no such restriction exists for halal.

The equipment for halal and vegetarian production must be thoroughly cleaned and can be used immediately after cleaning. For kosher, after thorough cleaning, the equipment might have to be left idle for a period as explained earlier in this chapter.

Insects are not kosher (except grasshopper), halal (except locust), or vegetarian. Insect by-products such as carmine and cochineal are not considered kosher by the majority of Orthodox rabbis, but some rabbis permit them. Some insect by-products are considered halal.

All plant materials are kosher and vegetarian; however, some restrictions exist in kosher regarding insect infestations and sources of certain alcoholic products, as explained earlier in this chapter. Plant materials are halal as long as they do not contain significant amount of alcohol or intoxicants.

The rules for halal and vegetarianism apply year round; however, for kosher, during Passover, additional rules might apply beyond everyday kosher.

Table 18.1 Comparative Summary of Kosher, Halal, and Vegetarian Guidelines

Description	Kosher	Halal	Vegetarian
Pork, pig, swine and carnivorous animals	Prohibited	Prohibited	Not applicable
Ruminants and poultry	Slaughtered by a trained Jew	Slaughtered by an adult Muslim	Not applicable
Blessing/invocation	Blessing before entering slaughtering area; not on each animal	Blessing on each animal while slaughtering	Not applicable
Slaughtering by hand	Mandatory	Preferred	Not applicable
Mechanical slaughtering	Not allowed	May be done for poultry under supervision	Not applicable
Stunning before slaying	Sometimes permitted	Permitted to render unconscious	Not applicable
Stunning after slaying	Permitted	May be permitted	Not applicable
Other restrictions about meat	Only front quarters used; soaking and salting required	Whole carcass used, no salting required	Not applicable
Blood of any animal	Prohibited	Prohibited	Not applicable
Fish	With scales only	Most accept all fish; some only with fish scales	Not applicable
Seafood	Not permitted	Varying degree of acceptance	Not accepted
Microbial enzymes	Accepted	Accepted	Accepted

	Kosher	Halal	Vegetarian
Biotech-derived enzymes	Accepted	Accepted	Accepted
Animal enzymes	Kosher-slaughtered only	Accepted sometimes	Sometimes accepted
Porcine enzymes	Maybe accepted	Not accepted	Sometimes accepted
Cattle gelatin	From kosher-slaughtered animals	From halal-slaughtered animals	Not applicable
Fish gelatin	Kosher fish only	Any fish	Not applicable
Pork gelatin	Allowed by liberal Orthodox rabbis	Not permitted	Not applicable
Dairy products, whey	Made with kosher enzymes	Made with halal enzymes	No restrictions
Addition of cheese culture	Must be added by a Jew	No restrictions	No restrictions
Alcohol	Permitted	Not permitted	Accepted
Combining meat and dairy	Not permitted	Not applicable	Not applicable
Insects and by-products	Grasshopper accepted; by-products not accepted	Locust and by-products accepted	By-products acceptable
Plant materials	All permitted	Intoxicants and alcohol not permitted	All acceptable
Sanitation of equipment	Cleaning; idle period required; kosherization/ritual cleaning	Thorough cleaning; no idle period required	Thorough cleaning
Special occasion	Additional restrictions during Passover	Same rules year-round	Same rules year-round

REFERENCES

Al-Qaradawi, Y. 1984. *The Lawful and Prohibited in Islam*, The Holy Quran Publishing House, Beirut, Lebanon.

Awan, J.A. 1988. Islamic food laws: philosophy of the prohibition of unlawful foods, *Sci. Technol. Islam. World*, 6(3), 151.

Awan, J.A. 1992. Islamic Codex Alimentarius, *Sci. Technol. Islam. World*, 10(1), 7-18.

Chaudry, M.M. 1992. Islamic food laws: philosophical basis and practical implications, *Food Technol.*, 46(10), 92-93, 104.

Chaudry, M.M. and Regenstein, J.M. 1994. Implications of biotechnology and genetic engineering for kosher and halal foods, *Trends Food Sci. Technol.* 5, 165-168.

Chaudry, M.M. and Regenstein, J.M. 2000. Muslim dietary laws: food processing and marketing, *Encyclop. Food Sci.*, 1682-1684.

Hussaini, M.M. and Sakr, A.H. 1983. *Islamic Dietary Laws and Practices*, Islamic Food and Nutrition Council of America, Bedford Park, IL.

Jackson, M. 2000. Getting religion — for your products, that is, *Food Technol.*, 54(7), 60-66.

Pickthall, M.M. 1994. Arabic text and English rendering of *The Glorious Quran*, Library of Islam, Kazi Publications, Chicago, IL.

Regenstein, J.M. and Chaudry, M. 2001. A brief introduction to some of the practical aspects of kosher and halal laws for the poultry industry, in *Poultry Meat Processing*, Sams, A.R., ed., CRC Press, Boca Raton, FL.

Regenstein, J.M. and Regenstein, C.E. 1979. An introduction to the kosher (dietary) laws for food scientists and food processors, *Food Technol.*, 33(1), 89-99.

Regenstein, J.M. and Regenstein, C.E. 1988. The kosher dietary laws and their implementation in the food industry, *Food Technol.*, 42(6), 86, 88-94.

Regenstein, J.M. and Regenstein, C.E. 2000. Kosher foods and food processing. *Encyclop. Food Sci.*, 1449-1453.

Riaz, M.N. 1996. Hailing halal, *Prep. Foods*, 165(12), 53-54.

Riaz, M.N. 1997. Alcohol: the myths and realties, in *A Handbook of Halaal and Haraam Products*, Vol. 2, Uddin, Z., Ed., Publication Center for American Muslim Research and Information, Richmond Hill, NY.

Riaz, M.N. 1999a. Halal food processing and marketing, in *10th World Congress of Food Science and Technology,* Book of Abstracts, Australian Institute of Food Science and Technology, Sydney, p. 44.

Riaz, M.N. 1999b. Examining the halal market, *Prep. Foods*, 168(10), 81-85.

Sakr, A.H. 1991. *Pork: Possible Reasons for its Prohibition*, Foundation for Islamic Knowledge, Lombard, IL.

The Vegan Society. 2002. Donald Watson House, U.K., www.vegansociety.com/html/products//.

The Vegetarian Society. 2002. Park Dale, U.K., www.vegsoc.org/info/definitions.html.

19

HOW TO GET HALAL CERTIFIED

As discussed earlier, there are ca. 1.3 billion Muslims in the world (Chaudry, 2002). Southeast Asia has 250 million Muslim halal consumers. Indonesia, Malaysia, Singapore, and many other countries have government mandates to import halal-certified products only. Recently, other countries in the region, such as Thailand and the Philippines, have initiated regulations to encourage both the export and import of halal products. In these countries, halal is considered as a symbol of quality and wholesomeness not only by Muslims but also by non-Muslims. The halal program was started in Malaysia during the early 1980s with the passage of the Halal/Haram Act and formulation of a high government level halal/haram committee. Under Malaysian regulations, meat and nonmeat food products have to be certified halal by a recognized halal authority. The Malaysian Department of Islamic Affairs has created two lists of approved entities for dealing with meat and poultry products exported to Malaysian from each country that wishes to export to Malaysia: (1) approved Islamic organizations, and (2) approved meat and poultry slaughterhouses or abattoirs.

The halal certification program started in Indonesia during the early 1990s and has developed into one of the strictest programs in the world. The program is executed by an organization known as LP-POM or the Assessment Institute for Food, Drugs and Cosmetics. This institution operates under the authority of the Indonesian Council of Ulama (ICU), generally known as Majelis Ulama Indonesia (MUI). The institute enjoys full backing of various government departments and agencies. Malaysia, Indonesia, Singapore, and Thailand have designed their own halal logo for the products offered for retail or food service. A company wishing to apply each country's logo must make an application to that country, even when the product is manufactured in another country. However, certificates and logos of other recognized agencies are acceptable as well.

North America has a Muslim population of 8 million: 7 million in the U.S. and 1 million in Canada (Cornell University Survey, 2002). Thus, there is a tremendous economic opportunity for companies to meet the needs of Muslims for halal food products. Perception of the word halal varies among different groups of Muslims. People from the Middle East associate halal with meat and poultry only, whereas persons of South Asian and Southeast Asian origin contend that all food and consumable products have to be halal. The latter group calls the meat and poultry slaughtered by Muslims as zabiha. Over the past 30 years, many halal markets, ethnic stores, and restaurants have sprung up, mainly in major metropolitan areas. The main group of products sold as halal or zabiha halal at these stores are fresh meat and poultry as well as imported ethnic products. The U.S food industry, for the most part, has ignored this population group and has concentrated its efforts toward export to Muslim nations. In many communities, Muslims working under local retailers have been slaughtering their own animals, and the concept of halal certification was foreign to them. In the late 1990s, however, small- to mid-size companies recognized the vacuum and the need to capture this niche. For retail and multilevel marketed products, halal products certified by just one of the certifiers (the Islamic Food and Nutrition Council of America) has increased from less than 200 over the past five years to over 1500 (Othman, 2003). Halal certification is now becoming popular for domestic products, as it has been for exported products (Riaz, 2002).

WHAT IS A HALAL CERTIFICATE?

A halal certificate is a document issued by an Islamic organization certifying that the products listed on it meet Islamic dietary guidelines, as defined by that certifying agency.

TYPES OF HALAL CERTIFICATES

■ Registration of a site certificate — this type of certificate signifies that a plant, production facility, food establishment, slaughterhouse, abattoir, or any establishment handling food has been inspected and approved to produce, distribute, or market halal food. This does not mean that all food products made or handled at such a facility are halal certified. A site certificate should not be used as a halal product certificate.

■ Halal certificate for a specific product for a specific duration — this type of certificate signifies that the listed product or products meet the halal guidelines formulated by the certifying organization. Such a certificate may be issued for a certain time period or for a specified quantity of the product destined for a particular distributor or importer. If the certificate is for a specific quantity, it may be called a batch certificate or a shipment certificate. Meat and poultry products, for which each batch or consignment has to be certified, generally receive a batch certificate.

■ Yearly certification — may be automatically renewed contingent on passing the annual inspection, through halal compliance and payment of the certification fee.

DURATION OF THE CERTIFICATE

The duration for which a certificate is valid depends on the type of product:

■ A batch certificate issued for each consignment is valid for as long as that specific batch or lot of the product is in the market, generally up to product expiration date or "Use By" date.

■ If a certified product is made according to a fixed formula, a certificate may be issued for a one-, two-, or three-year period. The product remains halal certified as long as it meets all the established and agreed-on production and marketing requirements. Often a system of occasional unannounced plant visits is used to confirm the plant's status.

WHO IS AUTHORIZED TO ISSUE HALAL CERTIFICATES?

Any individual Muslim, Islamic organization, or agency can issue a halal certificate, but the acceptability of the certificate depends on the country of import or the Muslim community served through such certification. For example, to issue a halal certificate for products exported to Malaysia and Indonesia, the body issuing the halal certificate must be listed on each country's approved list. More than 40 organizations issue halal certificates in the U.S., but only 5 have been approved by the Majelis Ulama Indonesia (MUI) (Table 19.1) and 16 by the Jabatan Kemajuan Islam Malaysia (JAKIM) (Table 19.2). Fifty percent of the certificates approved by JAKIM over the years are not even active in issuing halal certificates according to JAKIM sources.

The food industry not only needs to understand halal requirements for different countries and the principles of halal but also needs an understanding of the organizations which would best meet their needs — organizations

Table 19.1 U.S.-Based Islamic Organizations Approved by AIFDC-ICU, Indonesia, for Issuing Halal Certificates

Organization	Address	Tel./Fax Nos. and E-mail
International Institute of Islamic Thought c/q Marjac Abbatoir	555 Grove St. Herndon, VA 22970	703-421-1133
Islamic Center of Omaha	P.O. Box 4093 3511 North 73rd St. Omaha, NE 68104	402-571-0720
Islamic Food and Nutrition Council of America (IFANCA)	5901 N. Cicero Ave. Suite 309 Chicago, IL 60646	773-283-3708; 773-283-3973 (Fax); mchaudry@ifanca.org
Islamic Food Authority, Inc.	913 S. Shumaker Dr. Salisbury, MD 21804	410-548-1728; 410-548-2217 (Fax)
Islamic Services of America	P.O. Box 521 Cedar Rapids, IA 52404	319-362-1480; 319-366-4369 (Fax)

Source: From the Majelis Ulama Indonesia (MUI), 2001.

which can service their global needs and are acceptable to the countries of import.

Malaysia and Indonesia are the only countries that have a formal program to approve a halal-certifying organization. Other countries such as Saudi Arabia, Singapore, Kuwait, United Arab Emirates, Egypt, and Bahrain also do approvals of organizations for specific products or purposes. Some of the major voices in halal recognition worldwide include:

- Jabatan Kemajuan Islam Malaysia (JAKIM), Malaysia
- Majelis Ulama Indonesia (MUI), Indonesia
- Majlis Ugama Islam Singapura (MUIS), Singapore
- Muslim World League (MWL), Saudi Arabia

WHICH PRODUCTS CAN BE CERTIFIED?

Any product consumed by Muslims can be certified, whether the product is consumed internally or applied to the body externally. Products that are used as medicine do not require halal certification in many countries; however, knowledgeable consumers look for medicines that meet halal guidelines, and halal certification might be a good investment even for medicines.

Table 19.2 U.S.-Based Islamic Organizations Recognized by Malaysia for Issuing Halal Certificates

Organization	Address	Tel./Fax Nos. and E-mail
Ames Islamic Community Center	1221 Michigan Ave. Ames, IA 50010	515-292-3683
Fox Valley Islamic Center	201 Kenny St. Green Bay, WI 54301	414-336-5515
Halal Council Southeast Asia	913 South Schumaker Dr. Salisbury, MD 21804	
Halal Transaction, Inc.	P.O. Box 4546 Omaha, NE 68104	
Institute of Halal Food Control	4354 South Drezel Chicago, IL 60646	312-924-0646; 313-548-8841 (Fax)
International Institute of Islamic Thought	555 Grove St. Herndon, VA 22070	703-471-1133
Islamic Association Center and Mosque	P.O. Box 521 Cedar Rapids, IA 52406	319-362-0480
Islamic Center of Omaha	P.O. Box 4093 3511 North 73rd St. Omaha, NE 68104	402-571-0720
Islamic Food Authority of America	14203 Ballinger Terrace Burtonsville, MD 20866	301-890-1321; 301-434-1065 (Fax)
Islamic Society of Greater Harrisburg	223 W. Jackson St. York, PA 17403	717-846-6838
Masjid Fresno, The Islamic Center of Central California	2111 East Show Ave. Fresno, CA 93710	209-222-6686
Northern Virginia Islamic Center	P.O. Box 4336 Falls Church, VA 22044	703-941-6558
Islamic Food and Nutrition Council of America (IFANCA)	5901 N. Cicero Ave. Suite 309 Chicago, IL 60646	773-283-3708; 773-283-3973 (Fax); mchaudry@ifanca.org

Source: From the USDA, 2003.

HALAL CERTIFICATION PROCESS

The halal certification process starts with choosing an organization that meets one's needs for the markets to be serviced. If the target is a specific country, it is better to use an organization that is approved, recognized, or acceptable in that country. If the market areas are broader or even global, then an organization with an international scope would better meet one's needs.

The process starts with filling out an application explaining the production process, the products to be certified, and regions the products will be sold or marketed in, along with specific information about the component ingredients and manufacturing process and information about other products manufactured in the same facility. Most organizations review the information and set up an audit of the facility. At this time, it is advisable to negotiate the fees and clearly understand the costs involved.

During the review of the ingredient information or the facility audit, the organization might ask for replacement of any ingredients that do not meet its guidelines. Generally, the company and the halal-certifying agency sign a multiyear supervision agreement. Then a halal certificate can be issued for a specific shipment of a product or for a given period of a few months to several years. Overall, the process for halal certification of food products is not complicated, as explained in Figure 19.1.

STEPS INVOLVED IN HALAL CERTIFICATION

- Filling out an application to the organization on paper or on the Internet. Figure 19.2 shows a typical application.
- Review of the information by the organization, especially the type of product and its components.
- Inspection and approval of the facility. This includes review of the production equipment, inspection of ingredients, cleaning procedures, sanitation, and cross-contamination.
- For slaughterhouse, inspection involves review of holding areas, method of stunning, actual slaying, pre- and postslaughter handling, etc.
- Determining the cost and fees involved and signing of the contract.
- Payment of fees and expenses.
- Issuance of the halal certificate.

USE OF HALAL MARKINGS

When a product is certified halal, a symbol is normally printed on the package. For example, the Islamic Food and Nutrition Council of America

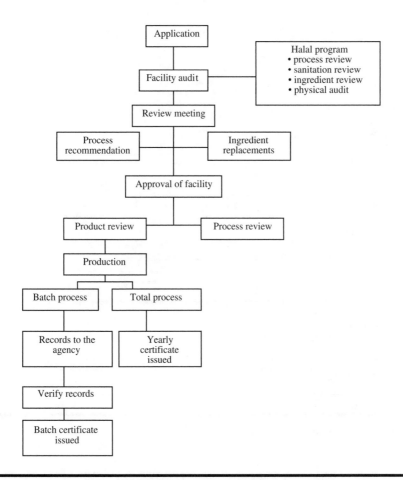

Figure 19.1   Flowchart of the halal certification process.

(IFANCA) uses the Crescent M symbol, which signifies "good for Muslims." Generic symbols, such as the word halal in Arabic inside a circle, are used by some companies with or without the endorsement of an organization. However, a product is better accepted by Muslim consumers if the logo signifies a reputable halal certification organization. Figure 19.3 shows a typical food with a Halal symbol. Figure 19.4 gives some of the halal markings and logos used by different countries and organizations.

Company Name:_____

Address:_____

_____
Tel. No.   Fax No.   Email Address   Website

Application Authorized By:_____
                           Signature   Date

Name and Title:_____

Product to be certified:_____

Type:_____

Geographic areas where product is/will be marketed. Please list all the countries.
USA____ Canada _____ Malaysia_____ Indonesia____Singapore____Saudi Arabia _____

____
UAE ____ Egypt _____ Pakistan _____ Others_____

Location of plant(s) where product is/will be manufactured: _____

_____
Contact person at plant_____
                        Name   Title

_____
Telephone Number   Fax Number   E-mail Address

LIST OF INGREDIENTS:
Please provide information about each ingredient, including specification sheets and suppliers.
Attach a separate sheet as needed.

_____
_____

Comments:_____

_____

Figure 19.2 Application for halal certification.

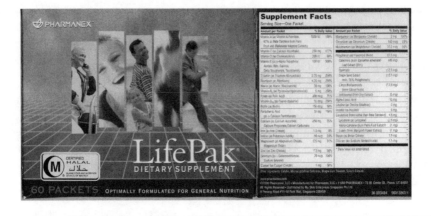

Figure 19.3 A dietary supplement with a halal symbol.

Figure 19.4 Halal markings and logos used by different countries and organizations.

REFERENCES

Chaudry, M.M. 2002. Halal certification process, presented at *Market Outlook: 2002 Conference*, Toward Efficient Egyptian Processed Food Export Industry in a Global Environment, Cairo, Egypt.

Cornell University Survey. April 2002. Study on American Muslim, Survey sponsored by Bridges TV, Orchard Park, NY.

Othman, R.M. 2003. Personal communication, rothman@ifanca.org and www.ifanca.org.

Majelis Ulama Indonesia. 2001. List of Islamic organizations approved by AIFDC-ICU, Mayid Istiglal, Jakarta, Indonesia, p. 1.

Riaz, M.N. 2002. Halal production standards and plant inspection requirements, paper presented at the *4th International Halal Food Conference on Current and Future Issues in Halal*, Toronto, Canada, April 21-23.

USDA. 2003. Export library. Eligible plants list. Malaysia, Islamic organizations recognized for issuance of halal certificates. www.fsis.usda.gov/ofo/export/lmalaysia.htm.

Appendix A

CODEX ALIMENTARIUS

Source: From www.fao.org/DOCREP/005/Y2770E/y277e08.htm.

GENERAL GUIDELINES FOR USE OF THE TERM HALAL

The Codex Alimentarius Commission accepts that there may be minor differences in opinion in the interpretation of lawful and unlawful animals and in the slaughter act, according to the different Islamic schools of thought. As such, these general guidelines are subjected to the interpretation of the appropriate authorities of the importing countries. However, the certificates granted by the religious authorities of the exporting country should be accepted in principle by the importing country, except when the latter provides justification for other specific requirements.

1. SCOPE

1.1 These guidelines recommend measures to be taken on the use of halal claims in food labeling.

1.2 These guidelines apply to the use of the term halal and equivalent terms in claims as defined in the "General Standard for the Labeling of Prepackaged Foods" and include its use in trademarks, brand names, and business names.

1.3 These guidelines are intended to supplement the "Draft Revision of the Codex General Guidelines on Claims" and do not supersede any prohibition contained therein.

2. DEFINITION

2.1 Halal food means food permitted under the Islamic law, and should fulfill the following conditions:

2.1.1 Does not consist of or contain anything which is considered to be unlawful according to Islamic law.

2.1.2 Has not been prepared, processed, transported or stored using any appliance or facility that was not free from anything unlawful according to Islamic law.

2.1.3 Has not in the course of preparation, processing, transportation, or storage been in direct contact with any food that fails to satisfy 2.1.1 and 2.1.2.

2.2 Notwithstanding Section 2.1:

2.2.1 Halal food can be prepared, processed, or stored in different sections or lines within the same premises where non-halal foods are produced, provided that necessary measures are taken to prevent any contact between halal and non-halal foods.*

2.2.2 Halal food can be prepared, processed, transported, or stored using facilities which have been previously used for non-halal foods, provided that proper cleaning procedures, according to Islamic requirements, have been observed.

3. CRITERIA FOR USE OF THE TERM HALAL

3.1 Lawful food — the term halal may be used for foods which are considered lawful. Under the Islamic law, all sources of food are lawful *except* the following sources, including their products and derivatives which are considered unlawful:

3.1.1 Food of animal origin
   a. Pigs and boars
   b. Dogs, snakes, and monkeys
   c. Carnivorous animals with claws and fangs such as lions, tigers, bears, and other similar animals
   d. Birds of prey with claws such as eagles, vultures, and other similar birds
   e. Pests such as rats, centipedes, scorpions. and other similar animals
   f. Animals forbidden to be killed in Islam, i.e., ants, bees, and woodpecker birds

---

* The draft guidelines were advanced to Step 8 subject to the advice of the executive committee on whether they fall outside the mandate of the commission and are contrary to the statements of principles adopted by the commission at its 21st session on the role of science in the Codex decision-making process and the extent to which other factors are taken into account.

     g. Animals which are considered repulsive generally, such as lice, flies, maggots, and other similar animals

     h. Animals that live both on land and in water such as frogs, crocodiles, and other similar animals

     i. Mules and domestic donkeys

     j. All poisonous and hazardous aquatic animals

     k. Any other animals not slaughtered according to Islamic law

     l. Blood.

3.1.2 Food of plant origin — intoxicating and hazardous plants except where the toxin or hazard can be eliminated during processing

3.1.3 Drink

     a. Alcoholic drinks

     b. All forms of intoxicating and hazardous drinks

3.1.4 Food additives — all food additives derived from Items 3.1.1, 3.1.2, and 3.1.3

3.2 Slaughtering — all lawful land animals should be slaughtered in compliance with the rules laid down in the "Codex Recommended Code of Hygienic Practice for Fresh Meat" and the following requirements:

3.2.1 The person should be a Muslim who is mentally sound and knowledgeable of the Islamic slaughtering procedures.

3.2.2 The animal to be slaughtered should be lawful according to Islamic law.

3.2.3 The animal to be slaughtered should be alive or deemed to be alive at the time of slaughtering.

3.2.4 The phrase Bismillah (in the name of Allah) should be invoked immediately before the slaughter of each animal.

3.2.5 The slaughtering device should be sharp and should not be lifted off the animal during the slaughter act.

3.2.6 The slaughter act should sever the trachea, esophagus, and main arteries and veins of the neck region.

3.3 Preparation, processing, packaging, transportation, and storage — all food should be prepared, processed, packaged, transported, and stored in such that they comply with Section 2.1 and Section 2.2 and the "Codex General Principles on Food Hygiene" and other relevant Codex standards.

4. ADDITIONAL LABELING REQUIREMENTS

4.1 When a claim is made that a food is halal, the word halal or equivalent terms should appear on the label.

4.2 In accordance with the "Draft Revision of the Codex General Guidelines on Claims," claims on halal should not be used in ways which could give rise to doubt about the safety of similar food or claims that halal foods are nutritionally superior to, or healthier than, other foods.

The Codex General Guidelines for the Use of the Term "Halal" were adopted by the Codex Alimentarius Commission at its 22nd Session, 1997. They have been sent to all Member Nations and Associate Members of FAO and WHO as an advisory text, and it is for individual governments to decide what use they wish to make of the Guidelines.

Appendix B

[HALAL INDUSTRIAL PRODUCTION STANDARDS]

Source: From Chaudry, M.M., Hussaini, M.M., Jackson, M.A., and Riaz, M.N. *Islamic Status for the Permissibility of Various Food Additives*, My Own Meals, Inc., Deerfield, IL, 3rd printing, September 1997. With permission.

CONTENTS

A. Introduction: The Goal and Purpose of Halal Supervision
B. Halal Supervision and Inspection
    1. Qualifications of Inspectors and Supervisors
    2. Training of Inspectors and Supervisors
    3. Requirements for the Supervising Organization
C. Meat and Poultry — Dhabiha Halal
    1. Species Acceptable (Halal) for Consumers
    2. Slaughtering Practices: A Humane and Spiritual Process
    3. Requirements for Inspectors, Supervisors, and Slaughtermen
    4. Unacceptable Practices
    5. Production Practices and Standards
    6. Acceptance of Meat Slaughtered by Jews or Christians
D. Fish and Seafood
    1. Species Acceptable as Halal
    2. Slaughtering Practices
    3. Standards and Requirements for Supervisors and Inspectors
E. Milk and Dairy Products such as Cheese
    1. Yogurt
    2. Cheese

Until recently, the word halal meant little to a producer who would pay a Muslim to bless the slaughter of animals in the plant, not use alcohol, and then put halal on the product to entice Muslim consumers. Standards were so lax that some producers used tape recordings of the prescribed blessings to "meet the minimum requirements." Other producers "read books" or learned from Muslims what the "rules" were and established production standards to self-certify the halal status of their products.

Many Muslims began to rely upon non-halal certified products, particularly meats, since halal certification had been virtually nonexistent. The Muslim community had no formal infrastructure established to monitor the authenticity of halal. Most halal-certified products were shipped overseas. Today, several Muslim organizations are policing U.S. producers to assure proper halal practices. U.S. companies failing to follow proper Muslim inspection standards may be destined for both domestic and international "black listings."

Many Muslim organizations reverted to stipulating that halal products had to be produced by Muslim-owned and operated companies, most of which were only state and not federally (U.S. Department of Agriculture) inspected. Products could not be sold across state lines. This created restraints of trade, making it impossible for institutions to comply with halal requirements. In addition, there were not enough trained and/or employed Muslim slaughtermen since the market was in its infancy and demand was fleeting.

My Own Meals, Inc., and J&M Company are very active in halal production. IFANCA is establishing halal production and supervising standards for various products. Together, we offer you this book as the first in a series of standards manuals to be issued as the market evolves in order to facilitate an understanding among producers of what halal standards mean and how to comply with them. Producers failing to meet these minimum standards will likely face challenges both domestically and internationally. The certification activity is one of the first segments of the halal infrastructure being created by Muslim organizations to assure that producers understand what is required to comply with halal standards.

This manual is from a producer's practical perspective, utilizing the halal standards IFANCA and others require. A company does not have to be owned, run, or operated by Muslims to produce halal products. However, any product destined for halal labeling must be produced under the strict supervision, assistance, counsel, and participation of trained and competent Muslim production inspectors. It is the company's responsibility to find trained Muslim inspectors and to comply with the requirements. It is the marketer's responsibility to assure the reputation of the Muslim supervising organization contracted to do the certification.

A. INTRODUCTION: THE GOAL AND PURPOSE OF HALAL SUPERVISION

The word halal means "proper and permitted." Halal food is permitted by Allah (God) for Muslim consumption. Dhabiha halal refers to meat and poultry properly slaughtered by a Muslim according to Islamic rituals. The word haram means "prohibited" or "forbidden" by Allah for Muslim consumption. Foods not prepared or processed using halal standards are forbidden to be consumed, just as would be alcohol or pork.

From a producer's perspective, these concepts are important to remember. Not only must the ingredient be reviewed for permissibility, but also suppliers of the ingredient must be Muslim approved as halal. This is because many ingredients may be from a variety of sources, including meat-based sources. Creating and approving a supplier network is time consuming, but may save a great deal of time later in the production and planning processes. Given the evolving market, it may be necessary and prudent to establish halal processes at a few key supplier locations, particularly those connected to meat/poultry and related ingredients.

B. HALAL SUPERVISION AND INSPECTION

Finding and hiring a reputable halal supervisory organization and properly trained onsite halal inspectors and supervisors are critical for current and future business dealings. Very few organizations meet these requirements. Look to these organizations as any other potential business partner. Evaluate what value they bring to the operation, including their reputation, which provides acceptability to the consumer.

In the past, many companies hired individual untrained local Muslims to supervise and certify their production practices as a means to meet the minimum requirement of having a Muslim associated with the production. This imprudent practice threatens the success and acceptability of production. There may be benefits in working with a local Islamic-certifying group by training them in specific production systems so that they may become a better supervisory organization.

Inspectors and supervisors require training not only in religious teachings, but also in production quality and control, product flow systems, cleaning, and the proper use of the production equipment. Training must include a working knowledge of ingredients and potential suppliers. For production locations running both non-halal and halal products, additional experience is required. Both the supervisory organization and the inspectors must have an experience base to properly establish procedures to assure that the cleaning and production control systems are adequate (including systems from receiving, storing, and producing through to end-item packaging).

B-1 Qualifications of Inspectors and Supervisors

Inspectors and supervisors must be Muslims, actively practicing their faith with a committed spirit and working knowledge of halal requirements as stipulated in the Quran and the Sunnah (teachings of Mohammed). Both must have demonstrated practical knowledge of food preparation according to halal practices and requirements. Such practical knowledge usually comes from experience in the production of food similar to that being produced. For example, experience requirements differ for restaurants, commercial kitchens for caterers, a USDA production facility, etc.

Supervisors are usually men. However, women trained and qualified to perform such functions are not specifically excluded from assuming these duties.

B-2 Training of Inspectors and Supervisors

Inspectors and supervisors must receive adequate training in each of the following areas, and such training must be documented. Training shall be provided by any or all of the following: the supervising organization, a credible organization hired to provide such training, or another supervising organization with which the inspectors and supervisors previously worked and whose training is approved by the supervising organization.

Some of the areas to be covered in training are:

a. Halal requirements as stipulated in the Quran and Sunnah
b. Working knowledge of the product(s) under its supervision
c. Production facility layout, management practices, and policies
d. Production processes for non-halal products (where appropriate) and for halal products and where and why differences exist
e. Machinery and production methods and standards (on-the-job under a trained supervisor is appropriate)
f. Labeling format(s), design(s), statements, etc. acceptable to the supervising organization
g. Handling and packaging aspects of the facility
h. Industry practices for that particular product(s) market
i. Professionalism and maintaining control over documents, processing areas, packaging, etc.
j. Maintaining of logs and records of production to assure only those items produced as halal are labeled as halal and only approved vendors and ingredients are used for production

B-3 Requirements for the Supervising Organization

The supervising organization is charged with the full responsibility for training and performance of its inspectors and supervisors, and must perform any and all testing of inspectors and supervisors to assure that they are adequately trained, knowledgeable of the halal requirements, and have demonstrated reliability, professionalism, and ability. The supervising organization arranges insurance and compensation considerations with the producer.

Every supervising organization must have a religious authority either on staff or on a consulting arrangement. The religious authority must be disclosed to the producer and to any consumer or customer requesting this information. This religious authority, together with the experience of the supervising organization, contributes to the credibility and reputation of the supervising organization. It also provides information as to preferred practices, particularly related to slaughtering procedures and ingredient approval. The supervising organization must maintain detailed records of:

a. Documented and approved production and packaging standards, procedures, and practices. Only such approved methods should be used by any facility.
b. Experience and training of inspectors and supervisors.

   c. Packaging materials and labels used by production facility.

   d. Surprise inspection visits and findings during such visits.

   e. Violations of accepted practices and policies and resolution.

   f. Control over use of organization's name and trademarks. Should the supervising organization decide to no longer certify the producer or any part or all of its products, the supervising organization must cancel authorization of the producer's right to use of its name and/or trademarks on those product(s). The organization must then take reasonable care to inform key customers or consumers that such items are no longer under its supervision as halal. When end users are consumers, such as through a restaurant or store, reasonable care should be taken to inform religious leaders in the area of such a change.

C. MEAT AND POULTRY — DHABIHA HALAL

C-1 Species Acceptable (Halal) for Consumption

Generally, goats, beef, lamb, rabbit, buffalo, deer, cattle, camels, and giraffes are acceptable animals for consumption. Permitted birds include turkey, chicken, fowl, hens, geese, and ducks.

The most noted exclusion of common meat sources is pork. Pork and all pork-derived ingredients come from swine. Swine is considered unfit for consumption by all Muslims and is haram. It is also haram to raise, trade, transport, or in any way derive benefit from pork or pork by-products.

Also excluded are all animals considered "beasts of prey having talons and fangs," such as lions, wolves, dogs, cats, tigers, hyenas, foxes, and jackals. Domesticated donkeys are also excluded. Excluded birds which "prey on the flesh of dead animals" include vultures, crows, eagles, falcons, pelicans, and other scavengers. In addition, the milk and eggs of prohibited species are similarly prohibited for consumption.

In all cases, any animal which is not properly slaughtered as discussed below may not be consumed. This includes improperly slaughtered animals as well as animals that die on their own from disease, altercations with other animals, or human cruelty.

C-2 Slaughtering Practices — A Humane and Spiritual Process

It is permitted to consume certain animals, provided the animal is given proper treatment and appreciation for its role in continued human existence. Care must be taken to reduce pain and suffering to a creature which will be slaughtered for sustenance.

At the moment of slaughter, the tasmiyah and takbir blessings are said over each animal or bird by a trained, religiously observant Muslim slaughterman. Saying a blessing after slaughter, solely at the beginning of the slaughter process, over the telephone, or on a prerecording are unacceptable practices. All products produced under these methods are unacceptable. Any representation that they may be halal is mislabeling. The blessings must only be uttered and supervised by a religiously observant Muslim in person.

C-3 Requirements for Inspectors, Supervisors, and Slaughtermen

Individuals assuming these roles must be educated and have experience in halal slaughtering practices. Any Muslim actively involved in slaughter must have experience in proper halal slaughter procedures. They must assure they use an extremely sharp knife and sharpen it often enough to keep it sharp. They must swiftly cut the esophagus, trachea, and jugular veins to rapidly release the blood and prevent pain or torture to the animal. The spinal column must not be cut, whether using mechanical or hand slaughter. The supervising organization is charged to assure that only properly trained inspectors, supervisors, and slaughtermen are permitted to participate in this phase of production.

C-4 Unacceptable Practices

Some producers label meat products as halal when these products are in fact not halal. It is possible that they may innocently believe they are following the "rules" to label the meat as halal. It is also possible that there is little concern other than marketing and sales driving these decisions.

When a meat product is labeled as halal with no reference to the certifying organization, the chances are high that the meat is mislabeled. A producer looking for halal meat as an ingredient must not assume that a meat item labeled as halal is indeed authentically halal. To be certain, request a halal certificate for each lot of meat to be used. Since meat is the most critical ingredient, the supervising organization must evaluate the supplier or recommend another supplier.

As advice to the producer buying halal meat as an ingredient, we caution you to take extra care in the area of meat. With every order, confirm the acceptability of the supplier. However, if your supervising organization states that there is only one acceptable meat supplier in the U.S., we recommend seeking an alternative supervising organization. This is an evolving supplier network today, surrounded by issues and problems. Be flexible and cautious as it evolves. If something does not seem right, it probably is not right. For the next few years, do not be surprised if you find your company together with your Muslim supervisory organization setting up halal standards at a supplier to get what you need.

Disallowed Slaughtering Practices

Under no circumstances is it permissible to:

- Say a blessing only at the start of the slaughtering process (but not throughout the process)
- Say a blessing only after all slaughtering is completed to cover all the animals slaughtered that day
- Use recordings of blessings to substitute for the devotion of an observant Muslim
- Accept the word of the slaughterhouse that humane methods were used and therefore should be considered the same as halal
- Accept that a product labeled as halal is indeed produced as halal
- Label a meat product as halal if there is no onsite Muslim participation
- Simultaneously process any pork or pork-derived product while producing halal-labeled meat
- Process any pork or pork-derived product immediately prior to the processing of any halal-labeled product without a full comprehensive and detailed cleaning (See Section N in production guidelines.)

C-5 Production Practices and Standards

It is imperative that the slaughter method used be done in a humane manner to avoid the infliction of unnecessary pain and suffering on the animal. The animal may not be mutilated, cut, or maimed while alive. It may not be bludgeoned. It is the responsibility of the slaughterer to take all due care to assure that the animal does not suffer.

The slaughter must be swift with a full frontal cut of the neck, throat (esophagus), and wind pipe (trachea). The two jugular veins in the neck must be cut without cutting the spinal cord. The blood of the animal must be thoroughly drained immediately upon slaughter. Only after the animal or bird is dead may the skin, head, feathers, and other parts be removed.

Each animal properly slaughtered according to halal standards is identified as Halal. Improperly slaughtered animals must not be marked halal. They must be removed and segregated from the halal meat. These carcasses may be sold as non-halal product. Documented and identifiable standards and procedures must be on-hand for each slaughter to assure the supervisor that the animals were not only properly prepared, slaughtered, and later handled for sale or distribution, but also to assure segregation of dhabiha halal animals from any animal, product, or by-product deemed non-halal. Should an animal die in the process of being slaughtered, either by stunning or other methods, prior to halal slaughter as required in this section, the animal is considered carrion and, therefore, haram.

Each package must be properly sealed under the supervision of a trained halal supervisor and properly marked as dhabiha halal only under his or her supervision. Any meat products which were not slaughtered under the above standards or not prepared and packaged as required by the halal supervisor may not be labeled as halal.

Methods of Slaughter

a. Slaughter at the Hand of a Human

When an animal is slaughtered by an individual, that individual must have received special training not only on the halal requirements but also in the methods and procedures required to cause no pain or suffering to the animal(s). This difficult task requires swiftness and strength. The faint of heart or weak must not be permitted to assume such a task. Otherwise, there is a strong probability that the animal will not be swiftly slaughtered. This could cause unnecessary pain and suffering to the animal. In this case, the requirements of halal slaughter may not be met.

Therefore, anyone not properly trained and strong enough to assume the responsibilities associated with the task of slaughter must not be permitted near animals. Should a supervising organization or anyone knowingly permit such a situation, everyone associated with the act will be responsible for the inhumane death of the animal. Specialized training and experience in this area cannot be taken lightly.

b. Mechanical or Rotary Knife Slaughter (Used for Birds)

Mechanical slaughter, or slaughter directly performed by machines, is somewhat controversial. Most Muslim countries now accept mechanically slaughtered birds if a trained Muslim working full-time at the machine utters the proper blessings. A machine is merely a tool to assist the slaughterman, much like a knife in the slaughterman's hand. Provided the machine properly cuts the esophagus, trachea, and two jugular veins, this mechanical slaughter can be used. Any birds missed by the machine must be slaughtered by the hand of the Muslim slaughterman. Contact IFANCA for a copy of *Mechanical Slaughtering Procedures*.

C-6 Acceptance of Meat Slaughtered by Jews or Christians

A slaughter performed by religiously observant Jews or Christians which properly meets all halal blessings and regulations may be used only under restricted and limited conditions. However, when meat slaughtered by Muslims is available, there is no need or reason to accept the slaughter of meat by Jews or Christians, and this exception provision does not exist.

Some Muslim families used to purchase meat and then bless it at home before cooking or eating it. In the strictest sense, the blessing must occur at the moment of slaughter, so reciting the blessings at home does not change the status of the meat.

D. FISH AND SEAFOOD

D-1 Species Acceptable as Halal

All fresh fish and seafood, including shellfish, are considered halal. Any processing or cooking of fish and products containing fish requires halal approval by a trained halal supervisor prior to being labeled as halal.

D-2 Slaughtering Practices

Fish does not need to be slaughtered. However, in no way shall a fish be permitted to suffer. It shall not be bludgeoned or its flesh cut from its body while alive. It shall not be cooked alive. It should be left to die by itself.

D-3 Standards and Requirements for Supervisors and Inspectors

Supervisors and inspectors shall be trained in the practices of fish production.

E. MILK AND DAIRY PRODUCTS SUCH AS CHEESE

E-1 Yogurt

Yogurt and yogurt products must not contain gelatin, unless that gelatin is determined to be from halal-slaughtered animals.

E-2 Cheese

Many cheeses contain rennet and other enzymes that are derived from animals. It is important to assure that these are derived from halal-approved animals or from microbial or plant sources. Other ingredients must also be from plant sources.

F. FRUITS AND VEGETABLES

Fresh fruits and vegetables are all considered halal. Processed fruits and vegetables may be unacceptable if they are produced in processing plants using non-halal oils, fats, preservatives, flavorings, colorings, etc. The procedures used in producing these items do not require onsite Muslim supervision. However, a prudent producer will work with its halal-supervising organization to ascertain that the supplier uses procedures to assure that the halal status of the fruits and vegetables has not been compromised.

In addition, ingredients or processing aids used in production or filling packages should be evaluated for halal status. Potentially problematic ingredients could be added flavorings, colorants, or preservatives (see Section I). Both the suppliers and the products must be approved by the halal-supervising organization. Also refer to Section L related to the halal status of packaging materials.

G. BREAD, BREADING, PASTRIES, AND CAKES

Bakery goods carry particular halal concerns. Breading on products such as fried chicken or cheese sticks or breading used as a "hidden" ingredient in stuffing or fillers may contain potentially questionable ingredients such as fats, oils, flavorings, colors, preservatives, or alcohol-based ingredients. It is important to assure that no alcohol-based or animal-based ingredients are used to ascertain halal status. Refer to the appendix for a listing of the halal status of many ingredients.

H. OILS AND FATS

Fats are rendered products which must be from halal-slaughtered animals. Haram preservatives or processing aids must not be used in vegetable-based oils.

I. ADDITIVES, COLORANTS, AND PRESERVATIVES

It is important to assure that the additives, colorants, and preservatives are from acceptable sources and are processed according to halal requirements without the use of alcohol-based carriers. Refer to the appendix for a listing of the halal status of many ingredients.

J. ALCOHOL AND ALCOHOL BY-PRODUCTS

Alcohol means ethyl alcohol. All products or ingredients containing alcohol are prohibited in Islam, even for cooking purposes or in fillings in candies.

Artificial and natural flavors, colors, and some types of meat or vegetable bases may contain alcohol products used to carry the flavor. They would be unacceptable to the requirements of halal. These may be hidden ingredients, so care must be exercised. The supervising organization may need to follow up on processes and ingredients. Note that vinegar, although a by-product or derivative of alcohol, is permitted in Islam. However, it is prudent to avoid the words "wine vinegar" in the label statements in order not to create confusion.

K. PREPARED FOODS — COMPLEX AND MULTIPLE INGREDIENTS

Prepared foods usually contain a wide variety of ingredients. As a result, the supervising organization takes extra care in reviewing and periodically approving supplier sources.

The production plant of complex and multiple ingredient meat-based products, such as soups, stews, and prepared meals, must have onsite Muslim supervision and participation. If the plant is producing vegetarian halal products, but usually produces non-halal meat-based products, the facility requires specific cleaning and halal preparation processes be performed prior to commencing halal production.

Extra Muslim attention and supervision is required for halal certification of the prepared foods produced. Onsite Muslim control and supervision is mandatory for any halal-certified product containing meat.

L. PACKAGING MATERIALS

Packaging materials are questionable in their halal status. While a plastic, microwavable container of frozen food may appear acceptable, the source of some of the ingredients used to create the plastics may be hidden. In many cases, stearates are used in the production of plastic containers. Stearates may be of animal origin.

Metal cans are also suspect. In many cases, the formation and cutting of the cans require the use of oils to aid in their manufacture. Such oils can also be animal-derived.

The supervising organization must investigate and evaluate packaging to determine its acceptability in order to approve food in that particular container. Often, a "melt-down" test is used to determine how much, if any, of the container contains haram ingredients. Producers may use only approved containers and suppliers.

M. RELIANCE UPON REPUTABLE, WELL-KNOWN, AND RESPECTED SUPERVISING ORGANIZATIONS

As halal certification standards evolve, supervisory organizations will be created. Those with consistent standards, organization, documented training, and standards will likely last and gain the respect of the Muslim community. Eventually, consumers, as well as foreign customers, will rely upon the certifying organization's standards and reputation to properly determine the halal status of products. Establishing strong, approved halal practices and standards in the production process will facilitate future developments in halal certification and acceptability.

N. GUIDELINES AND STANDARDS FOR PRODUCERS

N-1 Certification Process

For products to be properly labeled as halal, the production facility, along with its management, policies, and production practices, must be closely and carefully evaluated by a reputable supervisory organization. Where production practices must be revised to conform to halal standards, a trained Muslim, under the tutelage of a reputable supervisory organization, must assist in implementing the required production practices and changes.

a. For noncomplex or dedicated lines where the same processes are used everyday, such as in fruit canning plants, little, if any, Muslim supervision is required. Once production practices are documented and established, the trained supervisor does not need to supervise all aspects of production all the time. Future inspectors may arrive unannounced during a planned production day to test that the documented procedures are in place and being utilized. In this case, the plant and all production may be considered halal all the time, requiring only an annual review and certification letter.

b. For complex products or where halal and/or non-halal meat products are manufactured, the halal-certifying supervisory organization must be informed every time halal production is scheduled. Special cleaning and preparation of equipment and facility are necessary prior to commencing halal production. A trained Muslim inspector or supervisor must be physically on site at the commencement of production. He or she must identify and inspect ingredients to be used, confirm packaging and labeling, check and approve cleanliness of the facility and equipment to be used, and review halal procedures being followed. Sections N-3 to N-8 relate to requirements for complex production environments. While an annual certification letter may be issued to the producer, a control listing of lot codes produced under halal supervision must be maintained by the Muslim supervisory organization to attest to the halal status by lot code. Some customers require every batch be certified.

c. For slaughtering facilities, an onsite Muslim supervisor must be present during all phases of slaughter, segregation of halal meat, and labeling. Each slaughter is halal certified individually, unless the slaughterhouse is exclusively a halal-slaughtering facility.

N-2 Inspection and Approval of Plant and Production Processes

a. Overall plant and production approval — to approve a facility for general halal certification, it must consistently perform the same type of production and produce the same group of approved products using the same ingredients. The Muslim supervisory organization will:

■ Review the production facility layout, production procedures, policies, and practices. This will involve direct inspection and discussion with plant management and personnel.

■ Evaluate management reputation, integrity, and reliability.

■ Review and approve all ingredients and ingredient-supplier listings. Procedures must be established whereby all new suppliers or ingredients changed during the year's certification letter period are approved by the halal-certifying organization before being used.

■ Approve written procedures for halal-only certified production, which may differ from non-halal certified production.

■ Authorize (in writing) how and when the producer may use the Muslim organization's name and halal symbol on its labels or in advertising.

■ Make unannounced inspections of documentation and sign-in logs which document when a Muslim inspector was on site.

■ Review sanitation procedures and sanitation chemicals to evaluate adequacy of cleaning.

b. Plant approval where complex products are produced — complex products include those produced at a facility which produces meat-based halal products. Complex products also include nonmeat-based halal products where the facility also produces non-halal meat products for other customers at other times.

- There is to be an onsite Muslim inspection at the commencement of each production run for procedures, ingredients, cleanliness, and packaging/label review.
- Whenever a noncertified product is run between halal production runs, an onsite inspector must reinspect ingredients, cleanliness, and packaging.
- The supervisory organization must obtain a production report of what was produced as halal and non-halal during halal production periods, by lot code.
- Packaging control procedures must be in place to assure that only halal-certified production is labeled with halal certification. In this case, an inventory of packaging and labels before and after the production run must be made available upon request by the halal-certifying supervisory organization. Written agreements with packaging suppliers must be made whereby the halal supervisory organization approves all purchase orders for packaging materials where halal certification appears, prior to releasing any packaging materials to the producer.

COMPLEX PRODUCTION PROCESSING REQUIREMENTS

N-3 Preparing for Production

A complete and thorough cleaning must occur the day before commencement of halal production. Equipment, piping (CIP cleaned in place or dismantled), feed lines, conveyors, cooking equipment, utensils, buckets, stoves, ovens, kettles, retorts, dollies, totes, barrels, and all other equipment to be used for halal production must be thoroughly cleaned and free of foreign material. The inspector will touch and visually inspect. Any remaining food, grease, or other materials must be thoroughly cleaned and removed prior to production.

N-4 Documentation Required to Prove Halal Status of Ingredients

Producers supply the halal-supervising organization with a list of ingredients used for halal production together with all primary or secondary suppliers. A listing of ingredients and suppliers must also be provided to the inspector upon inspection for noncomplex products, and prior to commencing the production of complex halal products.

If the inspector finds that nonapproved alternative ingredients or unacceptable suppliers are being used, production and certification is immediately suspended or revoked unless and until cleared by the supervisory organization.

N-5 Segregation, Shipping, Storage of Ingredients and Packaging

Segregating ingredients to be used solely for halal production, such as halal meat, makes it easier for the halal-supervisory organization to rely upon management controls than onsite visual inspection of every aspect of production. Even though not required, this practice reduces the possibility of invalidating the halal production by using a nonapproved ingredient (especially meat and meat bases). The product can be sold, but not as halal, once the halal status is violated.

There is potential for significant negative ramifications if a non-halal product is boxed in error into a halal-certified box. If packaging inventory variances cannot be explained to the halal supervisory organization, all halal products produced become suspect, which may result in a suspension of halal certification of all halal products in inventory unless and until cleared.

Segregating sealed halal and non-halal products in shipping (i.e., LTL common carriers) is unnecessary. However, fresh non-halal meat, which can leak meat juices through packaging materials, should be segregated. If all packages are properly sealed to prevent cross-contamination, there is no need to segregate halal food products from non-halal food products in shipping.

N-6 Additional Requirements for Plants Using Pork, Lard, Pork-Derived Ingredients for Non-Halal Products

A special cleaning is required in facilities that prepare pork products during non-halal production periods. Under no circumstances may pork production occur simultaneously in the same room with halal-certified production. The plant should be clearly divided so that the pork production is segregated and produced away from the halal production. It may take place in the same facility, just not in the same segregated production section of the plant.

An onsite halal inspector must arrive prior to production and inspect the production area and equipment to be used for halal production which was previously used for pork production to be assured that everything is perfectly clean. In addition, the area and equipment must again be cleaned under Muslim supervision with steam, boiling water, or chemicals to thoroughly clean any possible remnants of pork or fat. This cleaning pertains to any surface which will touch halal-processed food, including fillers, kettles, utensils, mixers, etc.

N-7 Halal Supervisor's Control Required over Packaging

To assure that only properly prepared products carry halal certification, it is critical to implement controls over packaging and labels so that the manufacturer cannot label non-halal products as halal in error. The supervisory organization must have a written agreement between the company and its supplier(s) of printed packaging and labels that all purchase orders for such materials must carry the approval signature of the supervisory organization prior to shipping to the manufacturer.

The supervisory organization must then keep a reasonable accounting for packaging materials inventory compared with purchases and production (output). Any significant variations require a written explanation of the variance to gain assurance that no nonapproved production was labeled as halal.

Under no circumstances may a rubber stamp be used without the prior approval of the onsite inspector. In cases where the facility prints its own labels, a letter attesting to the number of labels printed must be signed by an officer of the company and supplied to the supervisory organization.

N-8 Labeling Requirements

Under no circumstances may a company put the supervisory organization's name and/or symbol on its packages as certified halal unless previously approved in writing by the supervisory organization. Once authorization to use the name and/or symbol has terminated by either time or through decertification or suspension, all materials containing the halal symbol must be destroyed.

All products labeled as halal should also carry the supervisory organization's name and/or symbol. Products labeled as halal without reference to the certifying organization's name will likely be those products initially under review by the Muslim community's new policing efforts.

Appendix B-1

Food Additive	Functions	Islamic Status
Acetic acid	Acidulant, preservative, flavoring agent	Halal
Adipic acid	Acidulant, flavoring agent	Halal
Agar	Gelling agent	Halal
Albumin	Protein fortifier, binder	Questionable[a]
Algin	Thickening, binding, and gelling agent	Halal
Annatto	Color	Halal
Artificial color/flavor	Adds color and/or flavor	Questionable[a,b]
Ascorbic acid	Vitamin C, dough conditioner, antioxidant	Halal

(continued)

Appendix B-1 (continued)

Food Additive	Functions	Islamic Status
Aspartame	Sweetener	Halal
Benzoic acid	Preservative	Halal
Beta carotene	Colorant, vitamin A	Halal
Butylated hydroxyanisole (BHA)	Antioxidant	Halal
Butylated hydroxytoluene (BHT)	Antioxidant	Halal
Caffeine	Beverage additive	Halal
Calcium caseinate	Protein fortifier, binding agent, whipping agent	Halal
Calcium propionate	Preservative, mold inhibitor	Halal
Calcium silicate	Anticaking agent	Halal
Calcium stearoyl lactylate	Whipping agent, dough conditioner, emulsifier	Questionable[a]
Calcium sulfate	Calcium source, filler, firming agent	Halal
Caramel	Colorant	Halal
Carrageenan	Stabilizer, gelling agent	Halal
Citric acid	Acidulant, antioxidant	Halal
Cornstarch	Thickener	Halal
Dextrin	Adhesive, flavor carrier	Halal
Diammonium phosphate	Leavening agent	Halal
Dipotassium phosphate	Emulsifier, buffering agent	Halal
Disodium guanylate/ inosinate	Flavor enhancers	Halal
EDTA	Sequestrant	Halal
Enzymes	Active additives	Questionable[c]
Erythorbic acid	Preservative	Halal
Ethyl alcohol	Extracting agent	Haram
Ferric orthophosphate	Food supplement	Halal
Ferrous fumarate	Food supplement	Halal
Ferrous gluconate	Food supplement	Halal
Fumaric acid	Acidulant	Halal
Gelatin	Gelling agent	Questionable[d]
Glutamic acid	Flavor enhancer	Questionable[a]
Glycerin	Humectant, crystallization modifier, plasticizer	Questionable[a]
Glyceryl monolaurate	Emulsifier	Questionable[a]

(continued)

Appendix B-1 (continued)

Food Additive	Functions	Islamic Status
Glycine	Dietary supplement, rancidity retardant	Questionable[a]
Guar gum	Stabilizer, thickening agent	Halal
Gum base	Gum	Questionable[a]
Hydrolyzed vegetable protein (HVP)	Flavor enhancer	Halal
Isopropyl citrate	Antioxidant	Halal
Lactic acid	Adjusts acidity, flavoring agent, preservative	Halal
Lactylated fatty acid esters	Emulsifiers	Questionable[a]
Lard	Hog fat	Haram
Lecithin (soy)	Emulsifier, dough stabilizer, anti-sticking agent, viscosity reducer, wetting agent	Halal
Magnesium stearate	Lubricant, binder, emulsifier, anticaking agent	Questionable[a]
Maltodextrin	Texturizer, crystallization inhibitor, bulking agent	Halal
Methylcellulose	Thickener	Halal
Methylsalicylate	Flavoring agent	Halal
Monoglyceride and diglyceride	Emulsifiers, dough conditioners, texture improvers	Questionable[a]
Monosodium glutamate (MSG)	Flavor enhancer	Halal
Oleoresins	Color and flavor additives	Questionable[a]
Oxystearin	Crystallization inhibitor, release agent	Questionable[a]
Pectin	Gelling agent	Halal
Phosphoric acid	Acidulant	Halal
Polysorbates	Emulsifiers	Questionable[a]
Potassium citrate	Sequestrant and buffering agent	Halal
Potassium sorbate	Preservative	Halal
Potassium stearate	Binder, emulsifier, plasticizer in chewing gum base	Questionable[a]
Propyl gallate	Antioxidant	Halal
Propylene glycol	Humectant, solvent	Halal
Propylene glycol alginate	Thickener, stabilizer, emulsifier	Halal

(continued)

Appendix B-1 (continued)

Food Additive	Functions	Islamic Status
Propylene glycol monostearate	Dispersing aid, crystal stabilizer, aeration increaser	Questionable[a]
Rennet	Milk coagulant used primarily in cheese	Questionable[c]
Saccharine	Sweetener	Halal
Shortening	Animal or vegetable fats or oils	Questionable[a]
Silicon dioxide	Anticaking	Halal
Sodium acid pyrophosphate	Leavening agent, preservative, sequestrant	Halal
Sodium alginate	Thickener, binder, gelling agent	Halal
Sodium ascorbate	Antioxidant, nutrient	Halal
Sodium benzoate	Preservative	Halal
Sodium bicarbonate	Leavening agent	Halal
Sodium caseinate	Protein fortifier, emulsifier, binding agent, whipping agent	Halal
Sodium citrate	Sequestrant	Halal
Sodium erythorbate	Antioxidant	Halal
Sodium hexametaphosphate	Binding agent	Halal
Sodium lauryl sulfate	Whipping agent, emulsifier	Questionable[a]
Sodium nitrate	Antimicrobial agent, preservative	Halal
Sodium nitrite	Antimicrobial agent, preservative	Halal
Sodium propionate	Preservative	Halal
Sodium silico aluminate	Anticaking and conditioning agent	Halal
Sodium stearate	Binder, emulsifier, anticaking agent	Questionable[a]
Sodium sulfite	Preservative	Halal
Sorbitan monostearate	Emulsifier, surfactant, dispersant	Questionable[a]
Sorbitol	Humectant	Halal
Stannous chloride	Antioxidant, preservative	Halal
Stearic acid	Lubricant, binder, defoamer	Questionable[a]
Stearoyl lactylate	Dough conditioner, emulsifier, whipping agent	Questionable[a]
Sulfur dioxide	Preservative, antimicrobial	Halal
Tallow	Animal fat	Questionable[a]
Tartaric acid	Flavor enhancer, acidulent	Halal
Titanium dioxide	Coloring agent	Halal
Vanilla	Flavorant	Questionable[b]

(continued)

Appendix B-1 (continued)

Food Additive	Functions	Islamic Status
Vanillin	Flavorant	Questionable[b]
Vinegar	Acidulant, flavorant	Halal
Whey	Source of lactose, milk solids, whey proteins	Questionable[c]
Yeast	Leavening and fermenting agent	Halal
Zein	Coating agent	Halal

[a] Source may be derived from, or contain ingredients of, animal origin.
[b] May contain ethyl alcohol which is haram.
[c] Rennet and other enzymes may come from non-halal slaughtered animals. Whey, a cheese by-product, may be derived from milk curdled with questionable rennet or enzymes.
[d] May be derived from pork. If gelatin is from fish or halal-slaughtered animals, it is halal.

AN INTRODUCTION TO MY OWN MEAL® AND J&M™ BRAND PRODUCTS

My Own Meals, Inc., and J&M Company produce and market halal-certified, refrigeration-free meals under the Islamic supervision and certification of IFANCA. All meat and poultry used are dhabiha halal certified. All meat-based meals are sold under the brand name of J&M™. All vegetarian meals are sold under the brand name of My Own Meal®.

Meals are packaged in 10-oz microwavable trays for home, office, school lunch, and institutional use, with an 18-month shelf life from date of manufacture. Meals are also available in 8-oz foil pouches with a 5-year shelf life from date of manufacture, primarily for campers, travelers, and military ration use. Refrigerated storage is not needed since the production process used is similar to the canning process. Other side dish items are being introduced. Meal varieties include:

- My Kind of Chicken® — chunks of chicken with brown rice, peas, and carrots in a light sauce; a favorite combination
- Chicken Mediterranean™ — chunks of chicken, tomatoes, potatoes, chick peas, and black olives in a tangy sauce; an enjoyable taste of the Mediterranean
- Chicken and Black Beans — chunks of chicken, black and kidney beans, tomatoes, potatoes, sweet peppers, and corn; a scrumptious meal with simple spicing

- Chicken and Noodles — chunks of chicken with kluski noodles, peas, and carrots in a light, deliciously seasoned sauce; a winning combination
- Chicken, Please® — a hearty stew with chunks of chicken, potatoes, corn, and carrots in a light sauce
- Beef Stew — a flavorful stew in a light sauce with chunks of lean beef, potatoes, sweet peppers, tomatoes, cabbage, zucchini, chick peas, and carrots
- Old World Stew — chunks of beef with brown rice, tomatoes, zucchini, and pinto beans; delicious with aromatic flavors
- Florentine Lasagna — small lasagna noodles with ricotta and parmesan cheeses, spinach, and pinto beans; a meal influenced by the tastes of Florence
- Cheese Tortellini — cheese-filled tortellini in a thick, seasoned tomato sauce flavored with parmesan cheese and pinto beans; perfect
- Vegetarian Stew — macaroni, vegetables, potatoes, barley, lentils, peanuts, and seasonings; a meal fit for vegetarian and hearty eater alike has a great taste and a variety of textures
- Pasta with Garden Vegetables — rotini pasta with a sweet pepper, mushroom, zucchini, and tomato vegetable medley; delicious and robust meal that is perfectly flavored with traditional Italian seasonings

All My Own Meal® and J&M™ products are available by mail order nationally by writing to:

> My Own Meals, Inc.
> Telephone: 847-948-1118
>
> J&M Company
> Telephone: 847-948-1290 (J&M)
> Fax: 847-948-0468
> P.O. Box 334
> Deerfield, IL 60015
> e-mail: myownmeals@worldnet.att.net

ABOUT THE AUTHORS

Dr. Muhammad Munir Chaudry

A food scientist, Chaudry left his executive technical position at Heller Seasonings to become president of IFANCA in 1984. A Ph.D. in food science, Chaudry has actively sought to improve halal production standards in the U.S. and abroad.

Mohammad Mazhar Hussaini

As a nutritionist who studied Muslim dietary life, Hussaini is the founding president and executive director of IFANCA. Hussaini continues to work with dietary concerns in his career and authored several books, including *Islamic Dietary Concepts & Practices* (IFANCA, 1993).

Mary Anne Jackson

A former Beatrice Foods Co. executive, CPA, and MBA, Jackson founded My Own Meals, Inc., in 1986. During her career at Beatrice since 1978, Jackson worked in accounting, finance, marketing, new product development, operations, and strategic and operations planning.

Dr. Mian Nadeem Riaz

A food research scientist at Texas A&M University in College Station, Texas, Riaz works with IFANCA to increase awareness of Muslim issues specifically related to halal certification. He evaluates the impact halal standards may have on food.

ABOUT THE ORGANIZATIONS

My Own Meals, Inc., and J&M Company

Founded in 1986 by Mary Anne Jackson, My Own Meals, Inc., pioneered and created the children's meals market by introducing its line of all-natural, refrigeration-free meals for children. In 1991, Jackson redirected its product and marketing focus to that of kosher and halal meals for both adults and children. In 1996, My Own Meals successfully introduced the first halal ration into the U.S. military ration system, which also included meat meals. The company continues its commitment and strategic emphasis on healthy, refrigeration-free meals. All J&M™ branded products are halal certified. All My Own Meal® branded products are kosher certified. All vegetarian My Own Meal® products are dual certified, both kosher and halal.

For more information, write to P.O. Box 334, Deerfield, IL 60015.

Islamic Food and Nutrition Council of America (IFANCA)

IFANCA was founded and registered in Illinois in 1982 by its board of directors with the objective of increasing awareness and understanding of Muslim dietary requirements, standards, and regulations. The organization is dedicated to scientific research in fields related to food, nutrition, and health. IFANCA provides halal supervision and certification for the production of qualifying products and producers. Its emphasis is on slaughtering, processed meats, foods, and kindred products.

For more information, write to P.O. Box 597722, Chicago, IL 60659-7722. *Halal Industrial Production Standards* is the first, single-source document describing halal production standards. Its goal is to define and create consistency in those standards and describe customer expectations of producers seeking halal labeling of products. As the demand for halal products develops, some of the standards may be clarified and modified over time, necessitating revisions to this document. Your Muslim supervisory organization will clarify and provide more detail to this book. However, only when a document exists can it be changed.

THE MARKET FOR HALAL FOOD PRODUCTS

The market for properly certified and labeled halal food products is in its infancy in the United States today. Domestic demand is limited. Halal production has been almost exclusively for exported meat products. For many reasons, including competition, international customers are becoming less accepting of halal labeling and requiring more information about the processes used. And the small domestic market is becoming vigilant and vocal in what it will and will not accept as it grows over the next decade.

Appendix C

MALAYSIAN GENERAL GUIDELINES ON THE SLAUGHTERING OF ANIMALS AND THE PREPARATION AND HANDLING OF HALAL FOOD

Source: From Jabatan Kenajuan Islam Malaysia. 2003. www.islam.gov.my//. With permission.

1. INTRODUCTION

These guidelines for the slaughtering of animals and the preparation and handling of halal food will have to be observed at all establishments involved in the processing of halal food. The guidelines will also serve as a basis for ascertaining the halal status of the establishments by the competent authority in Malaysia. These guidelines will apply to all foreign establishments intending to export their products to Malaysia and should be used together with existing guidelines on good manufacturing practices and hygienic sanitary requirements.

2. HALAL FOOD

Halal food means food permitted under the Islamic law (a law based on the al-Quran, al-hadith, ijma' (consensus) and qiyas (deduction of analogy) according to the syafei or any one of the hanafi, maliki, or hanbali school of thought or fatawa approved by the relevant Islamic authority and which fulfil the following conditions:

a. The food or its ingredients do not contain any components or products of animals that are not halal to Muslims according to the Syariah law or animals which are not slaughtered according to the Syariah law.
b. The food does not contain any ingredients that are considered najis (filthy) according to the Syariah law.
c. It is not prepared, processed, or manufactured using equipment that is contaminated with things that are considered najis according to the Syariah law.

During its preparation, processing, packaging, storage, or transportation, it should be physically separated from other food that does not meet the requirements stated in items (a), (b), or (c) or things that have been decreed as najis by the Syariah law.

3. HALAL DRINKS

All kinds of water and beverages are halal except those that are poisonous, intoxicating, hazardous to health, or mixed with najis.

3.1 NAJIS

According to the Syariah law, najis are:

3.1.1 Any liquid and objects discharged from all orifices of human beings or animals, such as urine, excrement, blood, vomit, pus including the sperm and ova of pigs and dogs except sperm and ova of all animals.
3.1.2 Dead animals or halal animals that are not slaughtered according to the Syariah law.
3.1.3 Halal food and drinks are considered najis if they are contaminated or have direct contact with things that are not permitted by Islam. The three kinds of najis (extreme, medium, or light) are applicable in this case.

3.2 FOOD AND DRINKS DERIVED FROM BIOTECHNOLOGY

Animals that have been treated (excluding feeding) with any product derived from non-halal sources (obtained through biotechnology or genetic engineering) become non-halal animals. Likewise, any food or drink derived from such animals is then deemed non-halal according to the Syariah law.

4. SOURCES OF HALAL FOOD AND DRINKS

4.1 ANIMALS

Animals can be divided into two categories.

4.1.1 Land Animals

All land animals are halal except the following:

4.1.1.1 Animals that are not slaughtered according to the Syariah law.

4.1.1.2 Pigs and dogs or their descendants.

4.1.1.3 Animals with long, pointed teeth (canines or tusks) which are used to kill preys such as tigers, bears, elephants, cats, and monkeys.

4.1.1.4 All predatory birds.

4.1.1.5 Animals that are decreed by Islam to be killed, such as mice, scorpions, crows, eagles, and mad dogs.

4.1.1.6 Animals that are forbidden to be killed, such as bees, ants, spiders, and woodpeckers.

4.1.1.7 Creatures that are considered dirty by the public, such as lice and flies.

4.1.1.8 Animals that live both on land and in water (amphibians), such as crocodiles, turtles, frogs, and seals.

4.1.1.9 Animals that have been treated (excluding feeding) with any product derived from non-halal sources.

4.1.2 Aquatic Animals

Aquatic animals are those which live in water and cannot survive outside it, such as fish. All aquatic animals are halal except those that are poisonous, intoxicating, or hazardous to health.

4.2 Plants

All type of plant products and plant derivatives are halal except those that are poisonous, intoxicating, or hazardous to health.

4.3 Microorganisms and Mushrooms

All types of microorganisms and mushrooms and their by-products and/or derivatives are halal except those that are poisonous, intoxicating, or hazardous to health.

5. HALAL SLAUGHTER

5.1 Preslaughter Conditions for Animals and Poultry

Only animals and poultry that fulfill the following will be allowed for slaughter:

5.1.1 All animals and poultry must be healthy, and free from any signs of wounds and disease or any form of disfigurement.

5.1.2 Animals and poultry should be treated in a humane manner prior to slaughter. Any ill-treatment, beating, or acts that may cause stress or fear are strictly forbidden on all animals and poultry awaiting slaughter.

5.1.3 Any acts of injury or cutting are prohibited on animals prior to their slaughter.

5.2 Conditions of Slaughter

The slaughtering of halal animals should be fully separated from the slaughter of non-halal animals. Halal slaughter should be carried out according to the following regulations:

5.2.1 Halal slaughter should be performed only by a Muslim (not a kitabi or others) who is of sound mind, mature, and who fully understands the fundamentals, rules, and conditions related to the slaughter of animals in Islam.

5.2.2 The animals to be slaughtered must be animals that are halal and can be eaten by a Muslim.

5.2.3 The animal must be fully alive or deemed to be alive at the time of slaughter.

5.2.4 The slaughtering should sever the trachea and esophagus. The carotid arteries and jugular veins will automatically be severed when both main vessels are cut off. The spinal cord should not be cut and the head is not to be severed completely. This is to bring about immediate and massive haemorrhage.

5.2.5 Slaughtering must be done only once. The "sawing action" of the slaughtering is permitted as long as the slaughtering implement is not lifted off the animal during the slaughtering. Any lifting of the knife is considered as the end of one act of slaughter. Multiple acts of slaughter on one animal are not allowed.

5.2.6 Bleeding must be spontaneous and complete.

5.2.7 Dressing of carcasses should only commence after ascertaining that the animal is dead.

5.2.8 Slaughtering implements, tools, and utensils must be utilized only for the slaughter of halal animals. The slaughtering implement or tool has to be kept sharp at all times.

5.2.9 The act of halal slaughter should begin with an incision on the neck at some point just before the glottis (Adam's apple) for animals with normal necks but after the glottis for animals with long necks, such as chicken, geese, turkeys, ostriches, and camels.

5.2.10 The phrase Bismillah (in the name of Allah) is highly encouraged to be immediately invoked before the slaughter of each animal. In certain mazhab (school of thought), this invoking of the phrase Bismillah is compulsory.

5.2.11 The objective of slaughtering is only for the sake of Allah and not for other purposes.

5.2.12 All halal verification certificates for halal meat must be issued, approved, and signed by an Islamic association which has been accepted by the government of Malaysia along with the health certificate from the exporting country. In the other words, the slaughterhouse must be under the supervision of the local Islamic authority, which is capable of auditing the halal certification, and duly recognized by the Department of Islamic Development Malaysia (DIDM).

5.3 Stunning of Animals

Stunning of animals prior to slaughter is permitted and should be in accordance with the following requirements:

5.3.1 Three types of stunners should be used to stun the animal, namely electrical, mechanical, or pneumatic stunner.

5.3.2 The use of the stunning equipment should be under the control of a Muslim supervisor or a trained Muslim slaughterman or halal certification authority at all times.

5.3.3 The animal should only be stunned temporarily. The stunning should not either kill or cause permanent injury to the animal.

5.3.4 Gadgets that are used to stun pigs will not be used to stun animals for halal slaughter.

5.4 Electrical Stunning

5.4.1 The electrical stunner should be of the type allowed by the government or Islamic authority in charge of slaughter.

5.4.2 A low and controlled voltage must be used so that the stunning will not damage the heart and brain or cause physical disability and death to the animals. The voltage of current used should be controlled by the halal certification authority.

5.4.3 The type of stunner used for slaughtering the halal animals should be "head only stunner" type, where both electrodes are placed on the head region.

5.4.4 Electrical stunning of poultry is allowed using water bath stunners only.

5.5 Mechanical Stunning

5.5.1 Mechanical stunners should only be used on cattle and buffaloes.

5.5.2 Nonpenetrative type (mushroom head) percussion stunner should be allowed.

5.5.3 The stunner should not penetrate or break the animal's head. Any injury, if caused, should not be a permanent injury.

5.5.4 The skull of the animal (after the head is skinned) should be checked and examined for permanent injury. If the skull is found to have been penetrated or broken, the carcass should be identified as non-halal and should be separated from halal carcasses.

5.6 Pneumatic Stunning

Pneumatic stunning or a stunner which uses air pressure is allowed for use in halal slaughter. It is a device operated by electrical power to produce strong air pressure and it does not contain any sharp objects. Air pressure directed toward the atlanto-occipital region will render the animal unconscious for a few seconds.

5.7 Mechanical Slaughter of Poultry

Slaughter of poultry by mechanical knife is permitted if it complies with the following procedures:

5.7.1 The operator (slaughterman) of the mechanical knife should be a Muslim. He will recite the prayer Bismillahir Rahmanir Rahim prior to switching on the mechanical knife and line.

5.7.2 The Muslim slaughterman who switches on the mechanical knife should be present in the slaughter area at all times (during slaughter).

5.7.3 In the event that the slaughterman has to leave the slaughter area, he should be replaced by another Muslim slaughterman. The former will stop the line and switch off the mechanical knife. To restart the operation the second slaughterman must recite the prayer as in 5.7.1, before switching on the mechanical knife and line.

5.7.4 The knife used should be of the single-blade type and must be kept sharp at all times.

5.7.5 During the act of slaughter, the throat, esophagus, and major blood vessels in the neck region (jugular vein and carotid artery) of the bird must be severed.

5.7.6 The slaughterman is required to check that each bird is properly slaughtered. He or other Muslim slaughterman should slaughter manually any birds that miss slaughter by the mechanical knife.

The birds should be dead as a result of slaughter before they are sent for scalding.

6. GENERAL REQUIREMENTS FOR HALAL MEAT PRODUCTS

6.1 Preparation, Processing, Packaging, Transportation, and Storage

All food (including meat) should be prepared, processed, packaged, transported, and stored in such a manner that it complies with all Islamic principles on halal food along with the "Codex General Principles on Food Hygiene" and other relevant Codex standards.

6.2 Processing and Handling

All processed food is halal if it meets the following conditions:

6.2.1 The product or its ingredients does not contain any components or products of animals that are not halal according to the Syariah law or animals that are not slaughtered according to the Syariah law.

6.2.2 The product does not contain anything in large or small quantities that is considered najis according to the Syariah law.

6.2.3 The product is prepared, processed, or manufactured, using equipment and facilities that are free from contamination with najis.

6.2.4 During its preparation, processing, packaging, storage, or transportation, the product should be totally separated from any food that does not meet all or any of the above three requirements or other things that are considered najis by the Syariah law.

6.3 Devices and Utensils

Premises and all facilities such as devices and utensils (machines) should only be used for processing halal food. The same premises and facilities are not allowed to be used for processing both halal and non-halal food, although they can be washed and cleaned properly as required by the Syariah law.

6.4 Storage, Display, and Serving

All halal products that are stored, displayed, sold, or served should be categorized and labeled halal at every stage so as to prevent it from being mixed or contaminated with things that are not halal.

6.5 Packaging and Labeling

The products should be properly packed and fulfil the following conditions:

6.5.1 Packaging materials should be halal in nature.

6.5.2 The packaging material should not contain any raw materials that are considered najis by the Syariah law.

6.5.3 The product should not be prepared, processed or manufactured by using equipment that is contaminated with things categorized as najis according to the Syariah law.

6.5.4 During its preparation, processing, packaging, storage, or transportation, the product should be totally separated from other food that does not meet all or any of the above three requirements or any other things that are considered najis by the Syariah law.

The packaging of meat and meat products should be carried out in a clean and hygienic manner in sound sanitary conditions. The word halal or equivalent terms should appear on the label of the product.

7. HYGIENE AND SANITATION

Hygiene has been given much emphasis in Islam and it includes various aspects of personal hygiene, clothing, equipment, and the premises used for processing or manufacturing of food. The objective is to ascertain that the food is produced under hygienic conditions and it is not hazardous to human health. The following conditions must be fulfilled:

7.1 The product should be processed and packed under strict hygienic conditions in premises licensed in accordance with good manufacturing practices.

7.2 The above principles are embodied in the document "Recommended International Code of Practice — General Principles of Food Hygiene (Ref. No. CAC/RCP 1-1969, Rev. 2 (1985))," and strict adherence to it is recommended. This document is in Section 1, Volume 1B, Codex Alimentarius — General Requirements (Food Hygiene).

8. COMPLIANCE

For a product to be deemed halal, it should comply with these guidelines, especially clauses 2 to 7. This should be verified through site inspection by a competent authority from Malaysia.

9. THE DEGREE OF CONFIDENCE IN HALAL CERTIFICATION

In the certification of the halal status, the examination process will cover all aspects of preparation, slaughtering, processing, handling, storage, transportation, cleaning, disinfection, and management practices. The examination should not create or give rise to any question or doubt. The examiner will only confer halal status when he is fully confident and is satisfied beyond any reasonable doubt on every aspect of the examination.

10. HALAL CERTIFICATION

10.1 The process of halal certification of a foreign establishment entails the site inspection of the plant by a competent authority from Malaysia along with the appointment of an Islamic organization that will be entrusted with the responsibility of supervising and monitoring the halal status at the plant concerned. The Islamic organization will have to be officially accredited by the DIDM. The halal certificate for products destined for export to Malaysia from this approved establishment will then be issued by the accredited Islamic organization.

10.2 The initial period of conferment of halal status to the approved establishment will be for a year. During this period the establishment will have to be monitored by the accredited Islamic organization. To maintain its halal status, the establishment will have to undergo periodic inspection every two years by the competent authority from Malaysia.

It is obligatory on the part of the Islamic organization to monitor any approved establishment and to submit an annual report on the halal status of the establishment to DIDM. Failure to submit such reports will result in both the withdrawal of the halal certification to the plant as well as the retraction of the recognition accorded to the Islamic organization.

11. CONFERMENT OF HALAL STATUS

All establishments that fulfil all the terms and conditions above can be conferred the accreditation as a halal establishment.

12. CANCELLATION/WITHDRAWAL OF HALAL STATUS

The Department of Islamic Development, Malaysia, reserves the right to cancel or withdraw the conferment of halal status to any establishment when the following occurs:

a. With the discovery of any breach or infringement in the halal requirements of the establishment concerned.

b. The failure to submit regular annual reports on the halal status of the establishment by the accredited Islamic organization.

c. Following the discovery that the Islamic organization is no longer functional or has been deregistered.

d. The Department of Islamic Development, Malaysia, will be not willing to entertain any requests seeking reinstatement of halal status subject to the following reasons:

　i. The failure to submit annual reports or the detection of any technical breaches will only warrant a repeat inspection of the establishment by the competent authority from Malaysia.

　ii. The detection of dishonest practices resulting in the loss of confidence by the competent authority in Malaysia will result in the suspension for a period not less than five years.

The suspended party can request for an inspection following the completion of the suspension period and which will be considered on its merits.

13. REQUESTS FOR INSPECTIONS FOR HALAL CERTIFICATION

The competent authority in Malaysia reserves the right to accept or refuse requests to perform inspections without the need to give any reason whatsoever in the event of such refusal.

Appendix D

SINGAPORE'S HALAL REGULATIONS AND IMPORT REQUIREMENTS

1. INTRODUCTION

1.1 Singapore's halal regulations and import requirements for food and food products, including their sale, are governed by three major legislations. These legislations are administered by three different government and quasi-government agencies.

1.2 The relevant legislation and the agencies that administer them are:

i. The Sale of Food Act 1973 and the Food Regulations 1988 — these regulations govern the importation of food or agricultural products into Singapore and they are administered by the Food Control Department of the Ministry of Environment.

ii. The Wholesome Meat and Fish Act 1999 — this act was recently passed by Parliament to control the import, export, and transshipment of meat products and fish products. It repeals the Slaughterhouses and Meat Processing Factories Act and is administered by the Primary Production Department of the Ministry of National Development.

iii. The Administration of Muslim Law Act (Amendment) Bill 1998 — this bill empowers the Islamic Religious Council of Singapore [or Majlis Ugama Islam Singapura (MUIS)] to regulate the halal certification of any product, service, or activity in Singapore.

2. OBJECTIVE

2.1 The objective of this paper is to briefly present the overall framework on Singapore's halal regulations and import requirements and focus on the more crucial aspects of the three major limbs of the relevant legislation.

3. IMPORT AND SALE OF FOOD PRODUCTS

3.1 The importation of food or agricultural products and their subsequent sale in Singapore are governed by the Sale of Food Act 1973 and the Food Regulations 1988. The Food Control Department (FCD) of the Ministry of the Environment administers these regulations.

3.2 To ensure the safety of all processed food being placed in the Singapore market, all food products, regardless of country of origin, are required to meet the standards laid down in the food regulations. Generally, imported food products must comply with the following requirements:

   i. The exporting food factory license certified by its government agency and photographs of the food factory, and delivery truck or chiller truck, to maintain food at certain temperatures, must be submitted to the FCD for verification.

Product Labeling

   ii. The food products are properly prepacked and labeled by the supplier in full accordance with the requirements as stipulated in the food regulations.

   iii. Product labels should bear the following basic information in English:

   a. An acceptable common name or description which is sufficient to indicate the true nature of the product.

   b. A statement listing all ingredients and additives used in the product in descending order by proportion of weight.

   c. If the food contains synthetic color, or tartrazine, it should be disclosed on the label.

   d. The minimum quantity of the food in the package expressed in terms of volumetric measure (for liquid food products) or net weight (for solid food products).

   e. Name of the country of origin.

   f. Name and address of the importer, distributor, or agent in Singapore.
   iv. The particulars stated in items (a) to (d) should be in printed letters not less than 1.5 mm in height.

Additional Labeling Requirements

   v. In addition to the above criteria, there are other labeling requirements for certain categories of food products such as:
      a. Milk and milk products.
      b. Special purpose foods.
      c. Food products that are claimed to be a source of energy, protein, vitamins, and minerals.
      d. Food products that contains artificial sweetening agents.
3.3 The labels for food products should not contain any claims that cannot by substantiated and claims that are prohibited under the food regulations.
3.4 Prepacked food products listed under the third schedule of the food regulations are required to be date-marked according to Regulation 10 of the food regulations.
3.5 Nutrition claims should include nutrition information panel in the form specified in the thirteenth schedule of the food regulations.

Food Import Registration

3.6 All prepacked foodstuffs imported for sale in Singapore must be registered with the FCD by submitting (mail or fax) the approved inward cargo clearance permit. Registration is free of charge.

Exemption

3.7 Food products are not required to be registered if they are for reexport, trade samples, diplomatic, personal use, or on transit.

Inspection and Sampling of Controlled Foodstuffs

3.8 After importing the following controlled foodstuffs, importers should contact the FCD for inspection and sampling of food before releasing for sale:
   i. Foodstuffs from Eastern Europe: check radioactive contaminants, namely, cesium 134 and cesium 137.

ii. Groundnut kernels, cashew nuts, and maize: check for afla-toxins.

iii. All starches and flours: check for aflatoxins and preservatives.

iv. Agar-agar and soybean products: check for boric acid.

v. Preserved fruits: check for cyclamate — aspartame, saccharin, acesulfame-K.

vi. Ceramic foodwares: check for lead and cadmium.

vii. Natural mineral, spring, and drinking water: for chemical and microbiological test, map indicating the source of water and letter of authenticity and health certificate.

viii. Whisky and brandy: original certificate of aged in wood for a period not less than 3 years from country of origin.

ix. Irradiated food requires a license from FCD.

4. IMPORT, EXPORT, AND TRANSHIPMENT OF MEAT AND FISH PRODUCTS

4.1 The Wholesome Meat and Fish Act 1999 was recently passed by the Parliament. This act aims to ensure that all meat and fish products are fit for human consumption and not diseased or contaminated.

4.2 This act repeals the Slaughterhouses and Meat Processing Factories Act and is administered by the Primary Production Department (PPD) of the Ministry of National Development. It is expected to be enforced by the second half of 1999.

4.3 Generally, this act will regulate the slaughtering of animals and the processing, packing, inspection, import, distribution, sale, trans-shipment, and export of meat and fish products. It will outline requirements too:

i. License all importers and exporters of meat and fish products.

ii. Regulate all aspects of in-bound and out-bound consignments of meat and fish products, i.e., to ensure that they meet with prescribed standards and comply with the packaging and labeling requirements.

iii. For purposes of exports, ensure that meat products are derived from authorized sources and that the consignment meets with the entry requirements of the destination country.

iv. License all slaughterhouses to ensure that all animals slaughtered are fit for human consumption, and carcasses of unfit animals are destroyed in the prescribed manner.

v. License all processing establishments such as factories and plants, and cold stores, including chillers, cold rooms, and freezers.

      vi.   License all wholesale markets selling any animals, or meat or fish products.

4.4 This act also allows for the enactment of subsidiary rules to further govern the specific provisions. The rules may provide for:

      i.   Inspection of premises and conveyances.

      ii.   Examination and certification of meat and fish products.

      iii.   Certification of meat products and fish products intended for export and the conditions under which such certificates may be withdrawn.

      iv.   Regulating or prohibiting the administration of drugs or other substances to animals before slaughter.

      v.   Regulating or prohibiting the acceptance at slaughterhouses of animals which are diseased or unfit for use.

      vi.   Controlling the processing and packing of meat or fish products in processing establishments.

      vii.   Prescribing requirements for the chilling, freezing, and storage of any meat product or fish product.

      viii.   Providing for the approval of any material used in packing.

      ix.   Regulating the construction, lighting, ventilation, air temperature, cleansing, drainage, water supply, maintenance, and good management of premises.

      x.   Regulating the construction, cleansing, and maintenance of all fixtures, appliances, instruments, utensils, and things used.

      xi.   Regulating the hygiene requirements to be observed by any person working therein so far as concerns the clothing, conduct, and health of such person.

      xii.   Controlling the application of chemicals, drugs, and other substances and the use of ingredients in the processing of meat or fish products.

      xiii.   Requiring information as to the nature and quantity of chemicals, drugs, or other substances which have been applied to any meat or fish product or of the ingredients which have been used in the processing of any meat or fish product to be shown on the labels of such products.

      xiv.   Prescribing the mode of dealing with animals and fish which are diseased or otherwise unfit for use in production.

      xv.   Requiring that any meat or fish product intended for export should conform to the requirements of the country to which it is to be exported, and prohibiting or restricting the export of any such product unless the prescribed requirements are complied with.

      xvi.   Providing for the handling and treatment of live animals intended for slaughter for human consumption.

xvii. Regulating methods for the slaughter of animals.

xviii. Providing for the standards of dressing animals slaughtered in licensed slaughterhouses.

xix. Regulating and controlling the import of animals, fish, meat, and fish products into Singapore and prescribing the mode of dealing with such animals, meat, and fish products upon their entry into Singapore.

xx. Prescribing the procedure for the taking and analysis of samples.

5. HALAL REGULATIONS

5.1 The Administration of Muslim Law Act (Amendment) Bill 1998 has tabled before Parliament. Due to be passed soon, this Bill will empower the Islamic Religious Council of Singapore (or MUIS) with the authority to regulate all aspects of halal certification in Singapore.

5.2 Prior to this Bill, MUIS could only rely on the Sale of Food Act. Offenders were taken to task under the ambit of false labeling. This reads: "No person shall sell any food which is labeled or advertised in a manner that is false, misleading or deceptive or is likely to create an erroneous impression regarding its value, merit or safety."

5.3 This bill will allow MUIS to "issue halal certificates in relation to any product, service or activity and regulate the holders of such certificates to ensure that the requirements of the Muslim law are complied with in the production, processing, marketing or display of that product, the provision of that service or the carrying out of that activity."

5.4 The bill will also allow MUIS to draw up a set of subsidiary rules to regulate the use and issue of halal certificates and the use of specified halal certification marks. It will also give MUIS the statutory powers to take recalcitrant offenders to task, including meting out penal measures.

6. BENEFITS OF BEING HALAL CERTIFIED

6.1 The halal certificate and the halal certification mark are authoritative, legitimate, and independent, and independent testimony to support one's claim as a manufacturer, distributor, or retailer that the food products meet halal requirements. For the discerning

Muslim consumer, they are a symbol of assurance and confidence that their Islamic needs are not compromised.

6.2 For exporters, the halal certificate may also boost the chances of meeting with the importing country's trade entry requirements. It enhances the product's marketability. The certificate is an internationally accepted means to ensure the halal status of foods or products.

7. GENERAL HALAL REQUIREMENTS

7.1 In order to ensure the effective control and administration of the halal certification scheme, all applicants must meet the terms and conditions of application before a certificate is issued. The main requirements are:

i.   All meat (including poultry) and meat products used must be slaughtered by qualified Muslims and sourced from a halal-certified vendor. Animal-based ingredients, such as gelatin and shortening, must also be derived from halal-certified sources. Pork, lard and any of its by-products, alcohol, and other non-halal foods are strictly prohibited from all ingredients.

ii. All imported meat (including poultry) and meat by-products must be halal certified by foreign Islamic authorities recognized and approved by MUIS.

iii. During the course of slaughtering, preparation, processing, transporting, or storage, all food products and ingredients must be clearly and visibly segregated and not come into contact with non-halal food products.

iv. There must be at least two Muslim employees of sufficient seniority involved in the procurement, verification, and production processes. Their role is vital to ascertain and continually ensure all aspects of the preparation processes are in compliance with Islamic requirements.

v.   The mixing of utensils and equipment of halal and non-halal food, which may occur during preparation, collection, washing, or storing, must be avoided. Separate washing and drying equipment/premises must be provided for. MUIS-appointed officials must ritually cleanse all utensils and equipment, including crockery, cutlery, dishwashers, chillers, freezers, and cold rooms if these have been used previously for preparing food with pork.

vi. To be a halal-certified restaurant or caterer, only halal food and beverages are to be served. Non-halal food, including pork and its by-products and alcohol, must not be served or used in any of the ingredients for the food and beverages. Employees are not allowed to bring any non-halal food or beverages onto the halal-designated premises.

vii. Fast-food chains with multiple outlets or restaurants operating under a franchise scheme must submit applications for all outlets.

viii. Applicants are advised to operate 100% halal and get their employees oriented and comfortable with the halal requirements set by MUIS before submitting their formal application.

8. AUDITS AND INSPECTIONS

8.1 Upon submission of application and in the course of being halal certified, random audits and surprise inspections are performed at the applicant's premises to ensure compliance with the halal requirements. Usually these audits achieve one or more of the following concerns:

i. Assess the seriousness of intention to go halal.

ii. Verify the authenticity of information and supporting evidence submitted.

iii. Assess the overall halal compliance and internal control systems.

iv. Assess the effectiveness and consistency of implementation.

v. Assess the role and the efficacy of the Muslim staff in guiding and ensuring compliance in the production process.

vi. Educate and reinforce understanding of halal requirements and compliance by the employees.

vii. Assess the risk of noncompliance.

viii. Report on areas for improvements.

9. IMPORT PROCEDURES

9.1 All newly imported foodstuffs have to be registered by the importer with the FCD. A copy of the inward declaration that has been approved by the Singapore Trade Development Board must be sent to the FCD by mail or fax. Singapore has introduced an electronic trade documentation system called TradeNet to facilitate import and export documentation (www.asianconnect.com/TDB/TRS/cmum.html).

9.2 Companies importing goods into Singapore need to contact the Trade Development Board for a central registration number. Registration covers a one-year period for a particular product. Items under import/export control may either require endorsement or license before the necessary licenses or approvals are obtained.

9.3 All shipments of meat-based and poultry products must be accompanied by export health certificates. A U.S. Department of Agriculture inspection certificate must be obtained for products that consist primarily of meat, i.e., sausages, franks, and patties. A state health certificate is also acceptable for those containing less than 3% meat. When shipping products that contain more than 3% meat, U.S. exporters should check with their local USDA office to see what type of certificate is required and how this document can be obtained. Inspection of meat and poultry shipments into Singapore is undertaken by the PPD. Importers of meat and poultry items must be registered with them.

Import Controls and Tariffs

9.4 Singapore imposes no quota restrictions and most goods may be imported under open license. Specific licenses are required for rice (excluding rice bran). There are no tariffs imposed on imported products except on tobacco, tobacco products, and alcoholic beverages.

Application for Cargo Clearance Permit

9.5 Applications for permits submitted through the TradeNet must bear the Ocean Unique Cargo Reference (OUCR)/House UCR (HUCR) Number for export consignments and Ocean Bill of Lading or the port of loading and the port of discharge (OBL)/House BL (HBL) Number for import consignments. These reference numbers are obtainable from the shipping agents, freight forwarders, and shippers.

10. TRADE ASSISTANCE AND CONTACTS

Majlis Ugama Islam Singapura
Islamic Centre of Singapore
273 Braddell Road
Singapore 579702
Tel.: 65-3591450
Fax: 65-3530363
www.muis.gov.sg

Food Control Department
Ministry of the Environment
Environment Building
40 Scotts Road
Singapore 228231
Tel: 65-7327733
Fax: 65-7319843 or 7319844

Custom & Excise Department
Customs Service Centre
Tel.: 65-3552000

Primary Production Department
Ministry of National Development
Maxwell Road #03-00
Singapore 069110
Tel.: 65-2221222
Fax: 65-2206068

Trade Development Board of Singapore
Manifest Audit Unit
Tel: 65-4334577 or 4334578 for general inquiries on the Manifest Reconciliation Statement Scheme

Port of Singapore Authority
Help Desk
Tel. 65-3211528 or 3211173

The Acts and Regulations described in this paper can be purchased at:

Singapore National Printers Corporation Ltd.
97 Ubi Avenue 4
Singapore 408754
Tel: 65-7412500
Fax: 65-7443770

Appendix E

NEW JERSEY HALAL FOOD LAWS

[FIRST REPRINT]

ASSEMBLY, NO. 1919

STATE OF NEW JERSEY

209TH LEGISLATURE

INTRODUCED MARCH 6, 2000

Sponsored by:
Assemblyman ALFRED E. STEELE
District 35 (Passaic)

SYNOPSIS

Prevents consumer fraud in the preparation, distribution, and sale of food represented as halal.

CURRENT VERSION OF TEXT

As reported by the Assembly Consumer Affairs and Regulated Professions Committee on May 22, 2000, with amendments.

An Act preventing consumer fraud in the preparation, distribution and sale of food represented as halal [1][amending the title and body of P.L.1988, c.154,][1] and supplementing P.L.1960, c.39(C.56:8-1 et seq.).

Be It Enacted by the Senate and General Assembly of the State of New Jersey:

1. (New section) Sections 1 through 6 of this act shall be known and may be cited as the "Halal Food Consumer Protection Act."

2. (New section) As used in this act:

   "Dealer" means any establishment that advertises, represents or holds itself out as selling, preparing or maintaining food as halal, including, but not be limited to, manufacturers, slaughterhouses, wholesalers, stores, restaurants, hotels, catering facilities, butcher shops, summer camps, bakeries, delicatessens, supermarkets, grocery stores, nursing homes, freezer dealers and food plan companies. These establishments may also sell, prepare or maintain food not represented as halal.

   "Director" means the Director of the Division of Consumer Affairs in the Department of Law and Public Safety or the director's designee.

   "Food" means a food, food product, food ingredient, dietary supplement or beverage.

3. (New section)

   a. Any dealer who prepares, distributes, sells or exposes for sale any food represented to be halal shall disclose the basis upon which that representation is made by posting the information required by the director, pursuant to regulations adopted pursuant to the authority provided in section 4 of P.L.1960, c.39(C.56:8-4), on a sign of a type and size specified by the director, in a conspicuous place upon the premises at which the food is sold or exposed for sale, as required by the director.

   b. It shall be an unlawful practice for any person to violate the requirements of Subsection (a) of this section.

4. (New section) Any person subject to the requirements of Section 3 of this act shall not have committed an unlawful practice if it can be shown by a preponderance of the evidence that the person relied in good faith upon the representations of a slaughterhouse, manufacturer, processor, packer or distributor of any food represented to be halal.

5. (New section) Possession by a dealer of any food not in conformance with the disclosure required by Section 3 of this act with respect to that food is presumptive evidence that the person is in possession of that food with the intent to sell.

6. (New section) Any dealer who prepares, distributes, sells or exposes for sale any food represented to be halal shall comply with all requirements of the director, including, but not limited to,

recordkeeping, labeling and filing, pursuant to regulations adopted pursuant to the authority provided in Section 4 of P.L.1960, c.39(C.56:8-4).

[1][7. The title of P.L.1988, c.154 is amended to read as follows:

An Act [to revise the kosher food law] concerning the preparation, distribution and sale of various foods, supplementing Chapter 21 of Title 2C of the New Jersey Statutes and repealing section 23B of P.L.1981, c.290. (cf: P.L.1988, c.154, title)][1]

[1][8. Section 1 of P.L.1988, c.154 (C.2C:21-7.2) is amended to read as follows:

1. As used in this act:
   a. "Advertise" means engaging in promotional activities including, but not limited to, newspaper, radio and television advertising; the distribution of fliers and circulars; and the display of window and interior signs.
   b. "Food," "food product," or "food commodity" means any food, food product or food preparation, whether raw or prepared for human consumption, and whether in a solid or liquid state, including, but not limited to, any meat, meat product or meat preparation; any milk, milk product or milk preparation; and any alcoholic or nonalcoholic beverage.
   c. "Food commodity in package form" means a food commodity put up or packaged in any manner in advance of sale in units suitable for retail sale and which is not intended for consumption at the point of manufacture.
   d. "Kosher" means prepared under and maintained in strict compliance with the laws and customs of the Orthodox Jewish religion and includes foods prepared for the festival of Passover and represented to be "kosher for Passover."
   e. "Halal" means prepared under and maintained in strict compliance with the laws and customs of the Islamic religion. (cf: P.L.1988, c.154, s.1)][1]

[1][9. Section 2 of P.L.1988, c.154 (C.2C:21-7.3) is amended to read as follows:

2. a. A false representation prohibited by this act shall include any oral or written statement that directly or indirectly tends to deceive or otherwise lead a reasonable individual to believe that a non-kosher food or food product is kosher.
   b. The presence of any non-kosher food or food product in any place of business that advertises or represents itself in any manner as selling, offering for sale, preparing or serving kosher food or food products only is presumptive evidence that the

person in possession offers the same for sale in violation of this act.

c. It shall be a complete defense to a prosecution under this act that the defendant relied in good faith upon the representations of a slaughterhouse, manufacturer, processor, packer or distributor, or any person or organization which certifies or represents any food or food product at issue to be kosher, kosher for Passover, or as having been prepared under or sanctioned by Orthodox Jewish religious requirements.

d. A false representation prohibited by P.L.1988, c.154 (C.2C:21-7.2 et seq.) shall include any oral or written statement that directly or indirectly tends to deceive or otherwise lead a reasonable individual to believe that a non-halal food or food product is halal.

e. The presence of any non-halal food or food product in any place of business that advertises or represents itself in any manner as selling, offering for sale, preparing or serving halal food or food products only is presumptive evidence that the person in possession offers the same for sale in violation of P.L.1998, c.154 (C.2C:21-7.2 et seq.).

f. It shall be a complete defense to a prosecution under P.L.1988, c.154 (C.2C:21-7.2 et seq.) that the defendant relied in good faith upon the representations of a slaughterhouse, manufacturer, processor, packer or distributor, or any person or organization which certifies or represents any food or food product at issue to be halal or as having been prepared under or sanctioned by Islamic religious requirements.

(cf: P.L.1988, c.154, s.2)][1]

[1][10. Section 3 of P.L.1988, c.154 (C.2C:21-7.4) is amended to read as follows:

3. A person commits a disorderly persons offense if in the course of business he:

a. (1) Falsely represents any food sold, prepared, served or offered for sale to be kosher or kosher for Passover;

(2) Removes or destroys, or causes to be removed or destroyed, the original means of identification affixed to food commodities to indicate that same are kosher or kosher for Passover, except that this paragraph shall not be construed to prevent the removal of the identification if the commodity is offered for sale as non-kosher; or

(3) Sells, disposes of or has in his possession for the purpose of resale as Kosher any food commodity to which a slaughterhouse plumba, mark, stamp, tag, brand, label or other means of identification has been fraudulently attached.

b. (1) Labels or identifies a food commodity in package form to be kosher or kosher for Passover or possesses such labels or means of identification, unless he is the manufacturer or packer of the food commodity in package form;

   (2) Labels or identifies an article of food not in package form to be kosher or kosher for Passover or possesses such labels or other means of identification, unless he is the manufacturer of the article of food;

   (3) Falsely labels any food commodity in package form as kosher or kosher for Passover by having or permitting to be inscribed on it, in any language, the words "Kosher" or "kosher for Passover," "parve," "glatt," or any other words or symbols which would tend to deceive or otherwise lead a reasonable individual to believe that the commodity is kosher or kosher for Passover; or

   (4) Labels any food commodity in package form by having or permitting to be inscribed on it the words "kosher-style," "kosher-type," "Jewish," or "Jewish-style," unless the product label also displays the word "non-kosher" in letters at least as large and in close proximity.

c. (1) Sells, offers for sale, prepares, or serves in or from the same place of business both unpackaged non-kosher food and unpackaged food he represents to be kosher unless he posts a window sign at the entrance of his establishment which states in block letters at least four inches in height: "Kosher and Non-Kosher Foods Sold Here," or "Kosher and Non-Kosher Foods Served Here," or a statement of similar import; or

   (2) Employs any Hebrew word or symbol in any advertising of any food offered for sale or place of business in which food is prepared, whether for on-premises or off-premises consumption, unless the advertisement also sets forth in conjunction therewith and in English, the words "We Sell Kosher Food Only," "We Sell Both Kosher and Non-Kosher Foods," or words of similar import, in letters of at least the same size as the characters used in Hebrew. For the purpose of this paragraph, "Hebrew symbol" means any Hebrew word, or letter, or any symbol, emblem, sign, insignia, or other mark that simulates a Hebrew word or letter.

d. (1) Displays for sale in the same show window or other location on or in his place of business, both unpackaged food represented to be kosher and unpackaged non-kosher food, unless he:

(a) displays over the kosher and non-kosher food signs that read, in clearly visible block letters, "kosher food" and "non-kosher food," respectively, or, as to the display of meat alone, "kosher meat" and "non-kosher meat," respectively;

(b) separates the kosher food products from the non-kosher food products by keeping the products in separate display cabinets, or by segregating kosher items from non-kosher items by use of clearly visible dividers; and

(c) slices or otherwise prepares the kosher food products for sale with utensils used solely for kosher food items;

(2) Prepares or serves any food as kosher whether for consumption in his place of business or elsewhere if in the same place of business he also prepares or serves non-kosher food, unless he:

(a) uses and maintains separate and distinctly labeled or marked dishes and utensils for each type of food; and

(b) includes in clearly visible block letters the statement "Kosher and Non-Kosher Foods Prepared and Sold Here" in each menu or sign used or posted on the premises or distributed or advertised off the premises;

(3) Sells or has in his possession for the purpose of resale as kosher any food commodity not having affixed thereto the original slaughterhouse plumba, mark, stamp, tag, brand, label or other means of identification employed to indicate that the food commodity is kosher or kosher for Passover; or

(4) Sells or offers for sale, as Kosher, any fresh meat or poultry that is identified as "soaked and salted," unless (a) the product has in fact been soaked and salted in a manner which makes it kosher; and (b) the product is marked "soaked and salted" on the package label or, if the product is not packaged, on a sign prominently displayed in conjunction with the product. For the purpose of this paragraph, "fresh meat or poultry" shall mean meat and poultry that has not been processed except for salting and soaking.

e. (1) Falsely represents any food sold, prepared, served or offered for sale to be halal;

(2) Removes or destroys, or causes to be removed or destroyed, the original means of identification affixed to food commodities to indicate that same are halal, except that this paragraph shall not be construed to prevent the removal

of the identification if the commodity is offered for sale as non-halal; or

(3) Sells, disposes of or has in his possession for the purpose of resale as halal any food commodity to which a slaughterhouse mark, stamp, tag, brand, label or other means of identification has been fraudulently attached.

f. (1) Labels or identifies a food commodity in package form to be halal or possesses such labels or means of identification, unless he is the manufacturer or packer of the food commodity in package form;

(2) Labels or identifies an article of food not in package form to be halal or possesses such labels or other means of identification, unless he is the manufacturer of the article of food;

(3) Falsely labels any food commodity in package form as halal by having or permitting to be inscribed on it, in any language, the words "halal" or "helal," or any other words or symbols, not limited to characters in Arabic writing, which would tend to deceive or otherwise lead a reasonable individual to believe that the commodity is halal; or

g. Sells, offers for sale, prepares, or serves in or from the same place of business both unpackaged non-halal food and unpackaged food he represents to be halal unless he posts a window sign at the entrance of his establishment which states in block letters at least four inches in height: "Halal and Non-Halal Foods Sold Here," or "Halal and Non-Halal Foods Served Here," or a statement of similar import.

h. (1) Sells or has in his possession for the purpose of resale as halal any food commodity not having affixed thereto the original slaughterhouse mark, stamp, tag, brand, label or other means of identification employed to indicate that the food commodity is halal; or

(2) Displays for sale in the same show window or other location on or in his place of business, both unpackaged food represented to be halal and unpackaged non-halal food unless he:

(a) displays over the halal and non-halal food signs that read, in clearly visible block letters, "halal food" and "non-halal food," respectively, or, as to the display of meat alone, "halal meat" and "non-halal meat," respectively;

(b) separates the halal food products from the non-halal food products by keeping the products in separate display cabinets, or by segregating halal items from non-Halal items by use of clearly visible dividers; and

(c) slices or otherwise prepares the halal food products for sale with utensils used solely for halal food items.

(cf: P.L.1988, c.154, s.3)][1]

[1][11.] 7.[1] This act shall take effect on the 180th day following enactment.

Appendix F

ILLINOIS HALAL FOOD LAWS

STATE OF ILLINOIS

92nd GENERAL ASSEMBLY

LEGISLATION

92 _SB0750enr
SB750 Enrolled   LRB9207942DJpr

An Act in relation to public health.

Be it enacted by the People of the State of Illinois, represented in the General Assembly:

Section 1. Short title. This Act may be cited as the Halal Food Act.

Section 5. Definitions. As used in this Act:

"Advertise" means to engage in promotional activities including, but not limited to, newspaper, radio, Internet and electronic media, and television advertising; the distribution of fliers and circulars; and the display of window and interior signs.

"Food," "food product," or "food commodity" means any food or food product inspected as required by law, or any food preparation from a source approved by the Department of Agriculture, whether raw or prepared for human consumption, and whether in a solid or liquid state, including, but not limited to, any meat, meat product or meat preparation; any milk, milk product or milk preparation; and any beverage.

"Food commodity in package form" means a food commodity put up or packaged in any manner in advance of sale in units suitable for retail sale and which is not intended for consumption at the point of manufacture.

"Halal" means prepared under and maintained in strict compliance with the laws and customs of the Islamic religion including but not limited to those laws and customs of zabiha/zabeeha (slaughtered according to appropriate Islamic code), and as expressed by reliable recognized Islamic entities and scholars.

Section 10. Deception prohibited.

a. It is a Class B misdemeanor for any person to make any oral or written statement that directly or indirectly tends to deceive or otherwise lead a reasonable individual to believe that a non-halal food or food product is halal.

b. The presence of any non-halal food or food product in any place of business that advertises or represents itself in any manner as selling, offering for sale, preparing, or serving halal food or food products only, is presumptive evidence that the person in possession offers the food or food product for sale in violation of subsection (a).

c. It shall be a complete defense to a prosecution under subsection (a) that the defendant relied in good faith upon the representations of an animals' farm, slaughterhouse, manufacturer, processor, packer, or distributor, or any person or organization which certifies or represents any food or food product at issue to be halal or as having been prepared under or sanctioned by Islamic religious requirements.

Section 15. Other offenses concerning halal food. It is a Class B misdemeanor for any person to:

1. Falsely represent any animal sold, grown, or offered for sale to be grown in a halal way to become food for human consumption;
2. Falsely represent any food sold, prepared, served, or offered for sale to be Halal;
3. Remove or destroy, or cause to be removed or destroyed, the original means of identification affixed to food commodities to indicate that the food commodities are halal, except that this

paragraph (3) may not be construed to prevent the removal of the identification if the commodity is offered for sale as non-halal;

4. Sell, dispose of, or have in his or her possession for the purpose of resale as halal any food commodity to which an animals' farm or slaughterhouse mark, stamp, tag, brand, label, or other means of identification has been fraudulently attached;

5. Label or identify a food commodity in package form to be halal or possess such labels or means of identification, unless he or she is the manufacturer or packer of the food commodity in package form;

6. Label or identify an article of food not in package form to be halal or possess such labels or other means of identification, unless he or she is the manufacturer of the article of food;

7. Falsely label any food commodity in package form as halal by having or permitting to be inscribed on it, in any language, the words "halal" or "helal," or any other words or symbols, not limited to characters in Arabic writing, which would tend to deceive or otherwise lead a reasonable individual to believe that the commodity is halal;

8. Sell, offer for sale, prepare, or serve in or from the same place of business both unpackaged non-halal food and unpackaged food he or she represents to be halal unless he or she posts a window sign at the entrance of his or her establishment which states in block letters at least 4 inches in height: "Halal and Non-Halal Foods Sold Here," or "Halal and Non-Halal Foods Served Here," or a statement of similar import;

9. Sell or have in his or her possession for the purpose of resale as halal any food commodity not having affixed thereto the original animals' farm or slaughterhouse mark, stamp, tag, brand, label, or other means of identification employed to indicate that the food commodity is halal; or

10. Display for sale, in the same show window or other location on or in his or her place of business, both unpackaged food represented to be halal and unpackaged non-halal food unless he or she:
    a. displays over the halal and non-halal food signs that read, in clearly visible block letters, "halal food" and "non-halal food," respectively, or, as to the display of meat alone, "halal meat" and "non-halal meat," respectively;
    b. separates the halal food products from the non-halal food products by keeping the products in separate display cabinets, or by segregating halal items from non-halal items by use of clearly visible dividers; and
    c. slices or otherwise prepares the halal food products for sale with utensils used solely for halal food items.

Section 20. Federal law. Nothing in this Act shall be construed to exempt halal food from any provisions of the federal Humane Methods of Slaughter Act of 1978 that may be applicable.

Section 90. The Consumer Fraud and Deceptive Business Practices Act is amended by adding Section 2KK as follows:

(815 ILCS 505/2KK new)

Sec. 2KK. Halal food; disclosure.

a. As used in this section:

"Dealer" means any establishment that advertises, represents, or holds itself out as growing animals in a halal way or selling, preparing, or maintaining food as halal, including, but not limited to, manufacturers, animals' farms, slaughterhouses, wholesalers, stores, restaurants, hotels, catering facilities, butcher shops, summer camps, bakeries, delicatessens, supermarkets, grocery stores, licensed health care facilities, freezer dealers, and food plan companies. These establishments may also sell, prepare or maintain food not represented as halal.

"Director" means the Director of Agriculture.

"Food" means an animal grown to become food for human consumption, a food, a food product, a food ingredient, a dietary supplement, or a beverage.

"Halal" means prepared under and maintained in strict compliance with the laws and customs of the Islamic religion including, but not limited to, those laws and customs of zabiha/zebeeha (slaughtered according to appropriate Islamic codes), and as expressed by reliable recognized Islamic entities and scholars.

b. Any dealer who grows animals represented to be grown in a halal way or who prepares, distributes, sells, or exposes for sale any food represented to be Halal shall disclose the basis upon which those representations are made by posting the information required by the Director, in accordance with rules adopted by the Director, on a sign of a type and size specified by the Director, in a conspicuous place upon the premises at which the food is sold or exposed for sale, as required by the Director.

c. Any person subject to the requirements of subsection (b) does not commit an unlawful practice if the person shows by a preponderance of the evidence that the person relied in good faith upon the representations of an animals' farm, slaughterhouse, manufacturer, processor, packer, or distributor of any food represented to be halal.

d. Possession by a dealer of any animal grown to become food for consumption or any food not in conformance with the disclosure required by subsection (b) with respect to that food is presumptive evidence that the person is in possession of that food with the intent to sell.

e. Any dealer who grows animals represented to be grown in a halal way or who prepares, distributes, sells, or exposes for sale any food represented to be halal shall comply with all requirements of the Director, including, but not limited to, recordkeeping, labeling and filing, in accordance with rules adopted by the Director.

f. Neither an animal represented to be grown in a halal way to become food for human consumption, nor a food commodity represented as halal, may be offered for sale by a dealer until the dealer has registered, with the Director, documenting information of the certifying Islamic entity specialized in halal food or the supervising Muslim inspector of halal food.

g. The Director shall adopt rules to carry out this Section in accordance with the Illinois Administrative Procedure Act.

h. It is an unlawful practice under this Act to violate this section or the rules adopted by the Director to carry out this section.

Appendix G

MINNESOTA HALAL
FOOD LAWS

MINNESOTA SESSION LAWS — 2001

Chapter 54-H.F.No. 149

An act relating to food; regulating the serving, selling, and labeling of certain religion-sanctioned food; amending Minnesota Statutes 2000, sections 31.59, subdivision 4; 31.661; proposing coding for new law in Minnesota Statutes, chapter 31.

Be It Enacted by the Legislature of the State of Minnesota:

Section 1. Minnesota Statutes 2000, section 31.59, subdivision 4, is amended to read:

Subd. 4. [Humane Methods] "Humane methods" means:

1. Any method of slaughtering livestock which normally causes animals to be rendered insensible to pain by a single blow of a mechanical instrument or shot of a firearm or by chemical or other means that are rapid and effective, before being shackled, hoisted, thrown, cast, or cut

2. The methods of preparation necessary to safe handling of the animals for halal ritual slaughter, Jewish ritual slaughter and of slaughtering required by the ritual of the Islamic or Jewish faith, whereby the animal suffers loss of consciousness by anemia of the brain caused by the simultaneous and instantaneous severance of the carotid arteries with a sharp instrument

Sec. 2. [31.658] [Halal Products]
Subdivision 1. [Halal Food Requirements] A person must not:

1. Serve, sell, or expose for sale food or food products, meat or meat products, or poultry or poultry products that are falsely represented as halal
2. Permit food, food products, meat or meat products, or poultry or poultry products, or the contents of a package or container to be labeled or inscribed with the "halal" sign unless the food or food products, meat or meat products, or poultry or poultry products have been prepared and maintained in compliance with the laws and customs of the Islamic religion
3. Make an oral or written statement that deceives or otherwise leads a reasonable person to believe that non-halal food or food products, meat or meat products, or poultry or poultry products are halal

Subd. 2. [Presumption] Possession of non-halal food or food products, meat or meat products, or poultry or poultry products in a place of business is presumptive evidence that the person in possession of them exposes them for sale.

Subd. 3. [Defense] It is a defense against a charge of misrepresenting non-halal food or food products, meat or meat products, or poultry or poultry products as halal that the person relied in good faith upon the representation of a slaughterhouse, manufacturer, processor, packer, distributor, or person or organization which certifies or represents a food or food product, meat or meat product, or poultry or poultry product as having been prepared under or sanctioned by Islamic religious requirements.

Sec. 3. Minnesota Statutes 2000, section 31.661, is amended to read:

31.661 [Marks, Stamps, Tags, Brands, or Labels]

No person shall:

1. Willfully mark, stamp, tag, brand, label, or in any other way or by any other means of identification, represent or cause to be marked, stamped, tagged, branded, labeled, or represented as kosher or as having been prepared in accordance with the orthodox Hebrew religious requirements food or food products not kosher or not so prepared
2. Willfully mark, stamp, tag, brand, label, or in any other way or by any other means of identification, represent or cause to be marked, stamped, tagged, branded, labeled, or represented as halal or as having been prepared in accordance with the Islamic religious

requirements, food or food products, meat or meat products, or poultry or poultry products not halal or not so prepared

3. Willfully remove, deface, obliterate, cover, alter, or destroy or cause to be removed, defaced, obliterated, covered, altered, or destroyed the original slaughterhouse plumba or any other mark, stamp, tag, brand, label, or any other means of identification affixed to foods or food products to indicate that such foods or food products are kosher or have been prepared in accordance with the orthodox Hebrew religious requirements

4. Willfully remove, deface, obliterate, cover, alter, or destroy or cause to be removed, defaced, obliterated, covered, altered, or destroyed the original halal sign, mark, stamp, tag, brand, label, or any other means of identification affixed to foods or food products, meat or meat products, or poultry or poultry products to indicate that the foods or food products, meat or meat products, or poultry or poultry products are halal or have been prepared in accordance with Islamic religious requirements

5. Knowingly sell, dispose of, or possess, for the purpose of resale to any person as kosher, any food or food products not having affixed thereto the original slaughterhouse plumba or any other mark, stamp, tag, brand, label, or other means of identification employed to indicate that such food or food products are kosher or have been prepared in accordance with the orthodox Hebrew religious requirements or any food or food products to which such plumba, mark, stamp, tag, brand, label, or other means of identification has or have been fraudulently affixed.

6. Knowingly sell, dispose of, or possess for the purpose of resale to any person as halal, any food or food products, meat or meat products, or poultry or poultry products not having affixed the original halal sign, mark, stamp, tag, brand, label, or other means of identification employed to indicate that the food or food products, meat or meat products, or poultry or poultry products are halal or have been prepared in accordance with Islamic religious requirements or any food or food products, meat or meat products, or poultry or poultry products to which the original halal mark, stamp, tag, brand, label, or other means of identification has been fraudulently affixed

Presented to the governor, April 26, 2001
Signed by the governor, April 30, 2001, 3:00 p.m.

Appendix H

CALIFORNIA HALAL FOOD LAWS

Bill Number: AB 1828 Chaptered

BILL TEXT

Chapter 102
Filed with Secretary of State July 2, 2002
Approved by Governor July 2, 2002
Passed the Senate June 20, 2002
Passed the Assembly April 22, 2002
Amended in Assembly February 25, 2002
Introduced by Assembly Member Bill Campbell (Coauthors: Assembly
    Members Aroner, Bates, Daucher, Harman, Leach, Maddox, and
    Robert Pacheco) (Coauthor: Senator Ackerman)

January 22, 2002
An Act to Add Section 383c to the Penal Code, Relating to Crime

LEGISLATIVE COUNSEL'S DIGEST

AB 1828, Bill Campbell. Halal Food.

Existing law provides that every person who, with intent to defraud, sells any meat or meat preparations falsely representing them to be kosher or prepared in compliance with Hebrew orthodox religious requirements, or who fails to indicate that both kosher and non-kosher meat is for sale in the same place of business, as specified, is punishable by a fine of not less than $100 nor more than $600, or imprisonment in a county jail for not less than 30 days nor more than 90 days, or by both that fine and imprisonment.

This bill would provide that a person who, with intent to defraud, sells any meat or meat preparations falsely representing them to be halal or prepared in compliance with Islamic religious requirements, or who fails to indicate that both halal and non-halal meat is for sale in the same place of business, as specified, is punishable by the same imprisonment and fine. By creating a new crime, this bill would impose a state-mandated local program.

The California Constitution requires the state to reimburse local agencies and school districts for certain costs mandated by the state. Statutory provisions establish procedures for making that reimbursement.

This bill would provide that no reimbursement is required by this act for a specified reason.

THE PEOPLE OF THE STATE OF CALIFORNIA DO ENACT AS FOLLOWS:

Section 1. Section 383c is added to the Penal Code, to read:

383c. Every person who with intent to defraud sells or exposes for sale any meat or meat preparations, and falsely represents the same to be halal, whether such meat or meat preparations be raw or prepared for human consumption, or as having been prepared under and from a product or products sanctioned by the Islamic religious requirements; or falsely represents any food product, or the contents of any package or container, to be so constituted and prepared, by having or permitting to be inscribed thereon the word "halal" in any language; or sells or exposes for sale in the same place of business both halal and non-halal meat or meat preparations, either raw or prepared for human consumption, who fails to indicate on his window signs in all display advertising in block letters at least four inches in height "Halal and Non-Halal Meats Sold Here;" or who exposes for sale in any show window or place of business as both halal and non-halal meat preparations, either raw or prepared for human consumption, who fails to display over each kind of meat or meat preparation so exposed a sign in block letters at least four inches in height, reading "Halal Meat" or "Non-Halal Meat" as the case may be; or sells or exposes for sale in any restaurant or any other place where food products are sold for consumption on the premises, any article of food or food preparations and falsely represents the same to be halal, or as having been prepared in accordance with the Islamic religious requirements; or sells or exposes for sale in such restaurant, or such other place, both halal and non-halal food or food preparations for consumption on the premises, not prepared in accordance with the Islamic ritual, or not sanctioned by Islamic religious requirements, and who fails to display on his window signs in all display advertising, in block letters at least four inches in height "Halal and Non-Halal Food Served Here" is guilty of a misdemeanor and upon conviction thereof be punishable by a fine of not less than one hundred dollars ($100), nor more than six hundred dollars ($600), or imprisonment in a county jail of not less than 30 days, nor more than 90 days, or both such fine and imprisonment.

The word "halal" is here defined to mean a strict compliance with every Islamic law and custom pertaining and relating to the killing of the animal or fowl from which the meat is taken or extracted, the dressing, treatment, and preparation thereof for human consumption, and the manufacture, production, treatment, and preparation of such other food or foods in connection wherewith Islamic laws and customs obtain and to the use of tools, implements, vessels, utensils, dishes, and containers that are used in connection with the killing of such animals and fowls and the dressing, preparation, production, manufacture, and treatment of such meats and other products, foods, and food stuffs.

Section 2. No reimbursement is required by this act pursuant to Section 6 of Article XIII B of the California Constitution because the only costs that may be incurred by a local agency or school district will be incurred because this act creates a new crime or infraction, eliminates a crime or infraction, or changes the penalty for a crime or infraction, within the meaning of Section 17556 of the Government Code, or changes the definition of a crime within the meaning of Section 6 of Article XIII B of the California Constitution.

Appendix I

MICHIGAN HALAL FOOD LAWS

Act No. 207
Public Acts of 2002
Approved by the Governor April 26, 2002
Filed with the Secretary of State April 29, 2002
Effective Date: April 29, 2002

STATE OF MICHIGAN

91st LEGISLATURE

REGULAR SESSION OF 2002

Introduced by Reps. Woronchak, Raczkowski, Pappageorge, Clark, Clarke, Cassis, and DeWeese

Reps. Anderson, Basham, Bob Brown, DeRossett, Hardman, Howell, Jacobs, Jamnick, Julian, Kolb, Lipsey, Mans, Mortimer, Murphy, Newell, O'Neil, Richardville, Rocca, Spade, Stewart, Toy, Van Woerkom, Vear and Zelenko named co-sponsors

ENROLLED HOUSE BILL NO. 5480

An Act to amend 1931 PA 328, entitled "An act to revise, consolidate, codify and add to the statutes relating to crimes; to define crimes and prescribe the penalties therefore; to provide for restitution under certain circumstances; to provide for the competency of evidence at the trial of persons accused of crime; to provide immunity from prosecution for certain witnesses appearing at such trials; and to repeal certain acts and parts of acts inconsistent with or contravening any of the provisions of this act," (MCL 750.1 to 750.568) by adding section 297f.

The People of the State of Michigan enact:

Sec. 297f.

(1) As used in this section, "halal" means prepared or processed in accordance with Islamic religious requirements.

(2) A person who, with intent to defraud, does any of the following is guilty of a misdemeanor:

a. Sells or exposes for sale in any place where food products are sold for consummation on or off the premises any meat, meat preparation, article of food, or food product, and falsely represents it to be halal, whether the meat, or meat preparation, article of food, or food product is raw or prepared for human consumption, either by direct statement orally, or in writing, which is reasonably calculated to deceive or lead a reasonable person to believe that a representation is being made that that food is halal.

b. Falsely represents any food product or the contents of any package or container to be so constituted and prepared, by having or permitting to be inscribed on the package or container the word "halal" in English.

c. Exposes for sale in any show window or place of business both halal and non-halal meat or meat preparations, or halal and non-halal food or food products, either raw or prepared for human consumption, and who fails to identify each kind of meat or meat preparation as "halal meat" or "halal food."

d. Displays on his or her window, door, or in his or her place of business, or in handbills or other printed matter distributed inside or outside of his or her place of business, words or letters in Arabic characters other than the word "halal," or any sign, emblem, insignia, symbol, or mark in simulation of same, without also displaying in English letters of at least the same size as such characters, signs, emblems, insignia, symbols, or marks, the words "We Sell Halal Meat and Food Only" or "We Sell Non-Halal Meat and Food Only," or "We Sell Both Halal and Non-Halal Meat and Food."

(3) Possession of non-halal food, in any place of business advertising the sale of halal food only, is presumptive evidence that the person in possession exposes the non-halal meat and food for sale with intent to defraud.

(4) A person who does any of the following is guilty of a misdemeanor:

   a. Willfully marks, stamps, tags, brands, labels, or in any other way or by any other means of identification represents or causes to be marked, stamped, tagged, branded, labeled, or represented as halal food or food products not halal or not so prepared.

   b. Willfully removes, defaces, obliterates, covers, alters, or destroys, or causes to be removed, defaced, obliterated, covered, altered, or destroyed the original slaughterhouse plumba or any other mark, stamp, tag, brand, label, or any other means of identification affixed to foods or food products to indicate that those foods or food products are halal.

   c. Knowingly sells, disposes of, or has in his or her possession, for the purpose of resale to any person as halal, any food or food products not having affixed to the food or food product the original slaughterhouse plumba or any other mark, stamp, tag, brand, label, or other means of identification employed to indicate that that food or food product is halal or any food or food products to which such plumba, mark, stamp, tag, brand, label, or other means of identification has been fraudulently affixed.

(5) The department of agriculture shall investigate and inspect the sale of halal food products and shall enforce this act. The department of agriculture may promulgate rules for the enforcement and administration of this section under the administrative procedures act of 1969, 1969PA 306, MCL 24.201 to 24.328.

This act is ordered to take immediate effect.

Clerk of the House of Representatives

Secretary of the Senate

Approved

Governor

Appendix J

EXPORT REQUIREMENTS FOR VARIOUS COUNTRIES

Source: From USDA. 2003. Library of Export Requirements, 8/5/2003.

EXPORT REQUIREMENTS FOR BAHRAIN

Eligible/Ineligible Products

A. Eligible products
   1. Fresh/frozen meat
   2. Poultry products

Slaughter Requirements

Ritual — Islamic halal slaughter requirements apply.

Labeling Requirements

A. For all products, storage temperature must be given with the refrigeration statement on the boxes to fully clarify the type of product being handled (e.g., "Keep frozen — store at or below ___ °C" or "Keep chilled (or refrigerated) — store between ___ °C and ___ °C").
B. In addition to the labeling features mandatory in the U.S., packaged meat and poultry must bear the following features:
   1. Bilingual labels — Arabic and English.
   2. Country of origin.

3. Production (slaughtering or freezing) and expiration dates are required only on shipping containers for institutional packaging.
   a. Date format requirements for Bahrain must conform to the following: day/month/year for products with a shelf life of 6 months or less and month/year for products with a shelf life of more than 6 months. Dating should be in numeric format and bilingual (English/Arabic).
   b. The expiration date is calculated from the date the product was frozen. The statement "Product was frozen 72 hours after slaughter" must be given in the "Remarks" section of FSIS Form 9060-5.
4. Use of the terminology "Keep refrigerated" is not acceptable on labels for frozen products.
5. Shelf life determination for a product must start from the production date.
6. Metric net weight.
7. Product identification.
8. If a certificate of Islamic slaughter is required, a statement that the animals have been slaughtered according to Islamic principles must be on the label.

C. The following methods of labeling are alternatives for adding production and expiration dates:
   1. Stickers, if used, must not obliterate label terminology and must be self-destructive on removal. Stick-on labels covering required labeling features are not permitted. No sticker carrying the production and/or expiration date is allowed on any product.
   2. Inserts must be accompanied by production and expiration dates. Inserts must be made of approved materials.

D. For prepackaged processed meat and poultry products, the production (packaging or freezing) and expiration dates and the net weights of frozen products are required on the label.

Documentation Requirements

A. FSIS Form 9060-5 should be obtained — correct production and expiration dates should be verified by inspection personnel prior to certification. All FSIS Form 9060-5 certificates must be dated and bear the signature and title of an FSIS veterinarian.

B. Islamic slaughter certification — a certification of Islamic (halal) slaughter is not mandatory. However, exporters should be aware that such a product will have limited distribution. U.S. exporters should contact the importer in Bahrain to determine whether a certificate of Islamic slaughter is required on a subject shipment. If required, the exporter must obtain a certificate of Islamic slaughter from an Islamic center or Islamic organization. A certificate of Islamic slaughter is a certificate issued by a member of a Muslim organization recognized by the importing country to provide this service; the certificate states that animals were slaughtered according to Muslim religious requirements. This certificate must be endorsed by a Bahrain Consul and must accompany all shipments.

C. For fresh/frozen products bearing halal labeling, the product must be accompanied by a certificate of Islamic slaughter or be accompanied by a written assurance from the exporter that an appropriate certificate will be supplied before the product reaches its destination.

D. For processed products, the product must be accompanied by a certificate of Islamic slaughter for the animals slaughtered to produce the raw materials used, or the product must be accompanied by a written assurance from the exporter that an appropriate certificate will be supplied before the product reaches its destination.

Handling/Storage Requirements

Bahrain requires that instructions for consumers concerning storage, preparation, and other special handling requirements accompany all shipments.

Other Requirements

A. Expiration period for fresh/frozen meat products — the period from slaughtering or freezing until arrival in Bahrain must not be more than 4 months. Product shall be maintained frozen at a temperature not higher than –18°C, with an expiration date of not longer than 12 months for beef and not longer than 9 months for minced meat and mutton.

B. Expiration period for fresh/frozen poultry products — the period elapsed from slaughtering or freezing until arrival in Bahrain must not be more than 3 months for frozen turkey, duck, goose, and chicken. The product shall be maintained frozen at a temperature not higher than –18°C, with an expiration date not longer than 12 months.

C. The requirements specified herein are to be used as guidelines only. It is the responsibility of the exporter to contact the importer to determine which requirements must be fulfilled for a particular shipment.

EXPORT REQUIREMENTS FOR EGYPT

Eligible/Ineligible Products

A. Eligible products
1. Fresh/frozen red meat and poultry
2. Red meat and poultry products
3. Horse intestines

Labeling Requirements

A. All products — storage temperature must be given with the refrigeration statement on the boxes to fully clarify the type of product being handled (e.g., "Keep frozen — store at or below __°C" or "Keep chilled (or refrigerate) — store between __°C and __°C")
B. Fresh/frozen and canned meat and poultry — in addition to the labeling features mandatory in the U.S., precut and packaged meat and poultry must bear the following features (in print):
1. Production (slaughtering or freezing) and expiration dates — name of month should be abbreviated or spelt out (e.g., Jan or January 1985). The expiration date is calculated from the date the product was first frozen. Calendar strips preprinted on label allowing the designation of calendar dates with literal translation are in frequent use.
2. Metric net weight — lettering and numbers for unit metric weight must be in Arabic.
3. Bilingual labels — all labels must be in Arabic and English.
C. Additional requirements — product information and the name and address of the importer must be printed in Arabic (with English optional) on a label inside the bag or wrapping (Cryovac) for fresh/frozen products, and for all products (additionally for fresh/frozen) this information must be on the immediate container and on boxes.

Certification Requirements

A. FSIS Form 9060-5 should be obtained. All FSIS Form 9060-5 certificates must be dated and have the signature and title of an FSIS veterinarian.
B. Special characteristics of product certification — exporters wishing to certify special characteristics of a product (such as types of pack or cut, weight range of product, or quality) to satisfy supplier–purchaser agreements or specifications can obtain such a certification

on a reimbursable basis from the grading services of the Agricultural Marketing Service, U.S. Department of Agriculture.

C. Additional certification

1. Beef products (including edible offals) — the following statements must be typed in the "Remarks" section of FSIS Form 9060-5: "Beef products, including edible offals, exported from the United States are derived from cattle that proved clinically or laboratory free of bovine immuno-deficiency virus (BIV) and enzootic bovine leukosis (EBL)" and "The United States is free of BSE. The meat is derived from cattle free of BSE and does not contain meat or animal protein originating from the United Kingdom."

2. Horse intestines — an FSIS Form 9060-10 should be obtained. The following statements are required in the "Remarks" section of the 9060-10: "The intestines are from animals free of contagious and infectious diseases" and "The United States is free of BSE."

Handling/Storage Requirements

Egypt requires that instructions for consumers concerning storage, preparation, and other special handling requirements accompany all shipments.

Other Requirements

A. Ship stores — Port Said is a free port. All U.S. products will be eligible for ship stores for any flagship.

B. Egypt import inspection — random laboratory samples for *Salmonella* are collected on meat and poultry product entering Egypt. Results of product testing by the country of origin prior to shipment are not honored by Egypt.

1. Beef (including beef livers) is accepted when 10% or less of the samples are positive.

2. Poultry is accepted when 20% or less of the samples are positive.

3. Exception to permissible levels of *Salmonella*. If the type of *Salmonella* identified is *S. typhi*, *S. paratyphi* A, B, and C, or *S. cholera-suis*, the shipment will be held pending a decision by the Ministry of Health.

C. Expiration period

1. Frozen meat (including beef and sheep livers with lymph nodes attached) must be shipped from the U.S. within 2 months of production date. The bill of lading will be used to confirm the date of shipping. The expiration period for frozen red meat

for direct consumption is 7 months from the date of freezing. All meat must arrive in Egypt with at least 50% of shelf life remaining.

2. Frozen poultry (including giblets) must be shipped from the U.S. within 2 months of the production date and arrive in Egypt within 3 months of the production date. The product must have an expiration date within 9 months of the production date.

D. When a frozen poultry sample is thawed, the amount of water collected must not exceed 5%.

E. Fat content of red meat — fat content of products for direct consumption must be no more than 7%. Fat content of red meat for processing must be no more than 20%.

Plants Eligible to Export

A. Beef liver — Egypt requires establishments exporting beef liver to be approved prior to shipment. The list of plants eligible to export beef liver to Egypt is found in the "Plant Lists" section of the FSIS Export Library. The FSIS Technical Service Center should be contacted at (402)-221-7400 for further information on new plant applications for approval to export beef liver to Egypt.

B. All other products — all federally inspected establishments are eligible to export to Egypt.

Information Regarding Halal Certification

The following is presented as information for the exporter. FSIS is not responsible for certifying that products intended for export to Muslim countries meet appropriate requirements for religious slaughter. Procedures for export certification do not include agency oversight of the halal process or review of the authenticity of the halal certificate.

Ritual Slaughter — Islamic Slaughter Certification

The exporter must obtain a certificate of Islamic slaughter from a member of an Islamic center or Islamic organization. A certificate of Islamic slaughter is a certificate issued by a member of a Muslim organization recognized by the importing country to provide this service. The certificate states that animals were slaughtered according to Muslim religious requirements. This certificate must accompany products labeled "halal." The certificate must be endorsed by the Arabian-American Chamber of Commerce or by an Egyptian Consulate and must accompany all shipments.

Halal Labeling

Halal-certified products must be labeled with statement that product has been slaughtered according to Islamic principles. (Halal labeling is an exporter and halal-organization responsibility.)

A. On fresh/frozen unprocessed products — products bearing halal label claims must be accompanied by an appropriate halal certificate.
B. On processed products with halal label claims — raw materials used in processed products with halal label claims must be accompanied by an appropriate halal certificate.

EXPORT REQUIREMENTS FOR INDONESIA

Eligible/Ineligible Products

A. Eligible products
1. Fresh/frozen red meat and red meat products.
2. Fresh/frozen poultry and poultry products.
B. Ineligible products — beef lungs are considered as "other edible offal of bovine animals" in Indonesia. They are eligible to be exported to Indonesia from countries that collect lungs as edible product. Lungs are considered inedible product in the U.S. and must be labeled "Not Intended for Human Food." Therefore, they are not eligible for export to Indonesia.

Documentation Requirements

Documentation requirements for fresh/frozen red meat and poultry products are as follows:

A. FSIS Form 9060-5 — Meat and Poultry Export Certificate of Wholesomeness— should be obtained.
1. Correct production and expiration dates should be typed in the "Remarks" section.
2. The form must be signed and dated by an FSIS veterinarian.
B. Poultry exported for further processing in Indonesia and subsequently exported to Japan must comply with the following statement:
1. Provided that the poultry complies, this statement may be provided in the "Remarks" section of FSIS Form 9060-5, if requested by the exporter: "There have been no outbreaks of

fowl pest (fowl plague) for at least 90 days in the United States. Further, in the area where the birds were produced (such an area being within a minimum radius of 50 kilometers from the production farm), Newcastle disease, fowl cholera, and other serious infectious fowl diseases as recognized by the government of the United States have not occurred for at least 90 days."

2. Poultry is restricted for export to Japan from certain states for specific periods. Certification must be provided that poultry did not originate from or transit through these states during the restricted periods relative to each affected state. Statements are obtainable from the Japan requirements for new certification or for recertifying previous shipments.

These statements are a requirement of the Japanese health officials for poultry produced in the U.S.

Plants Eligible to Export

A. All federally inspected U.S. meat and poultry establishments are eligible to export to Indonesia.

B. Sample shipments — shipments weighing less than 10 kg (22 lb) do not require an import permit.

Information Regarding Halal Certification

The following is presented as information for the exporter. FSIS is not responsible for certifying that products intended for export to Muslim countries meet appropriate requirements for religious slaughter. Procedures for export certification do not include agency oversight of the halal process or review of the authenticity of the halal certificate.

Ritual Slaughter — Islamic Slaughter Certification

The exporter must obtain a certificate of Islamic slaughter from a member of an Islamic center or Islamic organization. A certificate of Islamic slaughter is a certificate issued by a member of a Muslim organization recognized by the importing country to provide this service. The certificate states that animals were slaughtered according to Muslim religious requirements. This certificate must accompany products labeled "halal." The certificate must be endorsed by the Arabian-American Chamber of Commerce.

Halal Labeling

A. Halal labeling is an exporter and halal organization responsibility.
B. Frozen poultry and poultry products are not always required to have halal certification. Exporters are advised to verify this with their importer prior to shipping.
C. Processed products with halal label claims are not required to have a halal certificate, but the raw materials used in these processed products must be accompanied by an appropriate certificate.
D. Shipment for U.S. military personnel — the certificate of Islamic slaughter may be waived if products are shipped for consumption by U.S. personnel.

EXPORT REQUIREMENTS FOR IRAN

Eligible/Ineligible Products

A. Eligible products
   1. Red meat
   2. Poultry products

Slaughter Requirements

Ritual slaughter — poultry must be slaughtered by means of a sharp knife cutting through the skin, jugular vein, and trachea to result in thorough bleeding of the carcass in preparation for dressing and evisceration.

Documentation Requirements

A. Meat products — FSIS Form 9060-5 should be obtained.
B. Poultry products
   1. FSIS Form 9060-5 should be obtained.
   2. A USDA/FSIS letterhead certificate with the following statements should also be obtained: "The whole of the consignment described in the reference certificate was derived from poultry which:
      a. Were subjected at the slaughterhouse named in the certificate to ante-mortem inspection by an authorized veterinary officer and to post-mortem inspection under the supervision of an authorized veterinary officer and no sign of infectious disease was detected.

b. Originated from flocks under veterinary supervision in which, within the preceding two months, none of the following diseases have been diagnosed — Salmonellosis, Gumboro disease, pasteurellosis, Newcastle disease, fowl plaque, and ornithosis.

c. Have not been in contact at the slaughterhouse with any other poultry in which any of the diseases mentioned in (b) above have been diagnosed.

d. Were hatched, reared, and slaughtered in a state in which, after due inquiry, I am satisfied that no outbreak of a velogenic strain of Newcastle disease in commercial flocks has been recorded in the 6 months prior to slaughter.

e. For washing, chlorine has been added to carcass chill water according to USDA guidelines."

Note: Statement b and Statement d can be certified based on company certification stating the same statements signed by a company veterinarian.

3. USDA poultry grading — fresh (frozen), ready-to-cook broiler chickens (whole carcasses) must be accompanied by USDA grading certificate and meet the following requirements:

a. Broilers are Grade A, as shown by grading certificate and on cartons.

b. Weight of each broiler is within 850–1350 g (2–3 lb), averaging 1100 g (2.4 lb).

c. Birds have been slaughtered and frozen not more than 3 months before shipping, as shown on export certificate and by slaughter dates on cartons. Only first and last slaughter and freezing dates must be shown on the export certificate.

d. Each broiler is individually packed in airtight plastic material.

C. Special purchases — fresh (frozen) ready-to-cook poultry (whole birds) purchased under a Iranian government tender must meet all requirements specified in respective bids. Unless the tender lists conditions which must be certified by USDA, the inspector will only be concerned with normal reinspection for export and issuance of export certificate.

D. Special permit — the importer must have a permit issued by the Iranian Ministry of Agriculture.

E. Signatures on certificates — all required forms and supplementary statements must be dated and signed by an official veterinarian. Name and degree (DVM or equivalent) must be typed or printed after the signature.

Handling/Storage Requirements

For poultry, each broiler must be individually packed in airtight plastic material.

Plants Eligible to Export

Any federally inspected establishment is eligible to export to Iran.

EXPORT REQUIREMENTS FOR IRAQ

Eligible/Ineligible Products

A. Eligible products
1. Meat products — eligibility not determined at this time
2. Poultry products

Documentation Requirements

Certification requirements are as follows:

A. FSIS Form 9060-5 should be obtained.
B. Poultry — the government of Iraq purchases poultry products directly through U.S. exporters, submitting a tender for each shipment. The tender and resulting contract contain specifications which are certified by the poultry grading branch.

EXPORT REQUIREMENTS FOR JORDAN

Eligible/Ineligible Products

A. Eligible products
1. Fresh/frozen beef
2. Fresh/frozen poultry and poultry products

Documentation Requirements

Certification requirements are as follows:

A. Fresh/frozen beef — FSIS Form 9060-5 should be obtained.
B. Fresh/frozen poultry — FSIS Form 9060-5 should be obtained. The following statement must be typed in the "Remarks" section of FSIS Form 9060-5: "The United States is free of OIE List A poultry diseases."

Plants Eligible to Export

All federally inspected plants are eligible to export to Jordan.

EXPORT REQUIREMENTS FOR KUWAIT

Eligible/Ineligible Products

A. Eligible products
   1. Meat and meat products: beef, veal, lamb, mutton, and goat
   2. Poultry and poultry products
B. Ineligible products — pork and pork products

Processing Requirements

The following are processing requirements for meat and poultry:

A. Pork tissues or lard are not permitted in formulated products.
B. Products containing textured vegetable products are permitted in formulated products.

Labeling Requirements

A. Storage temperature must be given with the refrigeration statement on the boxes to fully clarify the type of product being handled. (e.g., "Keep frozen — store at or below __°C" or "Keep chilled (or refrigerated) — store between __°C and __°C")
B. Fresh/frozen and canned meat and poultry — in addition to the labeling features mandatory in the U.S., packaged meat and poultry must bear the following features:
   1. Country of origin.
   2. Statement that the product has been slaughtered according to Islamic principles.
   3. Bilingual labels — English and Arabic.
   4. The English section of the label should state the name of the product and the name and address of the manufacturer or the producer. Product names may deviate from the U.S. standards if approved according to 9 CFR 317.7 or 381.128. For example, chicken from the U.S. is commonly labeled "American Chicken."
   5. Metric net weight is required. Lettering and numbers for unit metric weight must be in Arabic. (*Note:* Gross weight is required for commercial invoices but is not required for product labels.)

6. Production (slaughtered or freezing) and expiration dates must be on individually packaged products. Date format requirements for Kuwait must conform to the following: day/month/year for products with a shelf life of 6 months or less and month/year for products with a shelf life of more than 6 months. Dating should be in numeric format and bilingual (English/Arabic).

C. The following methods of labeling are alternatives to meeting the requirements for labeling precut and packaged fresh/frozen meat and poultry:

1. Stickers may be used but must not obliterate label terminology and must be self-destructive on removal. Stick-on labels covering required label features are not permitted. No sticker carrying the production and/or expiration date is allowed on any product.

2. Inserts may be used but must be accompanied by production and expiration dates. Inserts must be made of approved materials.

Documentation Requirements

Certification requirements are as follows:

A. FSIS Form 9060-5 — Meat and Poultry Certificate of Wholesomeness — should be obtained. All FSIS Form 9060-5 certificates must be dated and have the signature and title of an FSIS veterinarian.

B. Islamic slaughter — in addition to FSIS certification, the exporter must obtain a certificate of Islamic slaughter from a member of an Islamic center or Islamic organization. A certificate of Islamic (halal) slaughter is a certificate issued by a member of a Muslim organization recognized by the importing country to provide this service. The certificate states that animals were slaughtered according to Muslim religious requirements. This certificate must accompany products labeled "halal." The certificate must be endorsed by the Arabian-American Chamber of Commerce or by a Kuwait Consulate and must accompany all shipments.

C. FSIS certification

1. Fresh/frozen unprocessed product — products bearing halal label claims must be accompanied by an appropriate halal certificate or a written assurance from the exporter that an appropriate halal certificate will be supplied to accompany that shipment before it reaches its destination. FSIS is not responsible for certifying that products intended for export to Muslim countries meet appropriate requirements for religious slaughter. Procedures for

export certification do not include agency oversight of the halal process or review of the authenticity of the halal certificate.

2. Raw materials used in processed products with halal label claims must be accompanied by an appropriate halal certificate.

D. Military shipments — the Kuwait Ministry of Foreign Affairs may exempt shipments consigned to the U.S. military from local import regulations. Without this exemption, the product will be subjected to inspection in accordance with local regulations. Each shipment consigned to the military should be accompanied by FSIS Form 9060-5 and a letter from the U.S. Embassy in Kuwait stating that the shipment is for use by the military.

Handling/Storage Requirements

A. Product requiring special handling — Kuwait requires that instructions for consumers concerning storage, preparation, and other special handling requirements accompany all shipments.

B. Packaging — all fresh/frozen products must be visible through the wrapper.

C. Product arrival and expiration dates
1. All imported packaged products with a shelf life of more than one year must arrive in Kuwait within 6 months from the production date on the label.
2. All imported products with a shelf life of one year or less must arrive within the first half of the shelf life or within 3 months from the production date, whichever is less.

Plants Eligible to Export

All federally inspected meat or poultry plants are eligible to export to Kuwait.

EXPORT REQUIREMENTS FOR MALAYSIA (PENINSULAR, EAST, WHICH INCLUDES STATES OF SABAH AND SARAWAK)

Eligible/Ineligible Products

Eligible products:

A. Meat products
1. Fresh/frozen beef, veal, and lamb/mutton carcass, primal cuts, as well as processed products from Malaysian-approved meat plants.

2. Meat and meat by-products, e.g., livers, spleens, hearts, brains, and other edible parts are eligible provided halal identity is maintained.
3. Beef lungs are eligible provided:
   a. The lungs must originate from establishments approved for export to Malaysia and must be accompanied by a halal certificate issued by an approved Islamic center.
   b. The lungs must be undenatured and labeled according to Section 325.8 of the regulations, i.e., "Beef lungs — not intended for human food."
   c. The only USDA certification which must be issued is a dated USDA letterhead indicating weight, number of cartons, title of MPI veterinarian signing the letterhead certificate (e.g., DVM or Veterinary Medical Officer), which specifies the following statements:
      i. "The lungs originate from animals that passed ante-mortem and post-mortem inspection and were inspected as follows: anterior, middle and posterior mediastinal, and right and left bronchial lymph nodes were incised and the curved surfaces of the lungs were palpated."
      ii. "Foot-and-mouth disease has not existed since 1929, and rinderpest has never existed in the United States."
   d. All federally inspected pork products are eligible to be exported.
B. Poultry products — fresh/frozen poultry and poultry products from Malaysian-approved plants.

Slaughter Requirements

Slaughter must be performed without stunning; however, use of mushroom stunning devices is acceptable provided the brain is not penetrated. (Animals will be rejected if brain is penetrated.)

Labeling Requirements

Beef lungs must be undenatured and labeled according to Section 325.8 of the MPI regulations, i.e., "Beef Lungs — Not Intended for Human Food."

Documentation Requirements

A. Permit requirements — exporters must obtain a permit through the importer. A permit is issued by the Malaysian Department of Veterinary Services, permitting the importation of meat and meat by-products and poultry and poultry by-products into Malaysia. (It is not necessary for the USDA inspector to verify the permit.)

B. Certification requirements
1. FSIS Form 9060-5 should be obtained. All FSIS Form 9060-5 certificates must:
   a. Be dated and have the signature and title of an FSIS veterinarian with degree and title (such as DVM or equivalent degree) printed or typed after the signature.
   b. Be accompanied by a veterinary certificate on USDA letterhead stating:
      i. "Foot-and-mouth disease has not existed since 1929, and rinderpest has never existed in the United States."
      ii. "The meat covered by this certificate originated from animals slaughtered in Est. __ (No.)" [This statement is required only for bovine and ovine meat and meat by-products which require a certificate of Islamic (halal) slaughter.]
      iii. "Swine fever (hog cholera) has not existed in the United States since 1978." (This statement is required only for pork and pork products.)
2. Correct production and expiration dates must be verified by inspection prior to certification.
3. Poultry products must contain the following statement in the remarks section of FSIS Form 9060-5: "The (poultry) products were derived from (poultry) subject to ante-mortem and post-mortem examinations and have been found to be free from infectious and contagious disease. The (poultry) products are fit for human consumption, and every precaution has been taken to prevent contamination prior to export. Foot-and-mouth disease has not existed since 1929, and rinderpest has never existed in the United States."

C. Certificate of Islamic laughter
1. In addition to FSIS certification, the exporters must obtain a certificate of Islamic (halal) slaughter from a member of an approved Islamic center or Islamic organization. A certificate of Islamic slaughter is a certificate issued by a member of a Muslim organization recognized by the importing country to provide this service. The certificate states that animals were slaughtered according to Muslim religious requirements. This certificate must accompany products labeled "halal." The certificate must be endorsed by an approved Islamic center.

2. Pork products do not require a certificate of Islamic (halal) slaughter.
3. Poultry/poultry products may be imported without halal certification, but may not be marketed as halal.

Other Requirements

Halal meat must be maintained separate and apart from non-halal meat. Product shipped from a slaughter plant to a processing plant must be identified, segregated, and accompanied by an Islamic slaughter (halal) certificate. The slaughter plant management furnishing halal meat to processing plants is responsible for informing the plant recipients that the meat must be kept identified and segregated from non-halal meat.

A. Bovine, ovine, porcine, and poultry slaughter and processing establishments and cold storage facilities are subject to inspection and approval by Malaysia. A list of establishments approved by Malaysia via this process is available in the FSIS Export Library. Malaysia may also allow the entry of product from plants not on the approved list. On the request of the exporter, export certification may be issued for products originating or stored in federal establishments not on the approved list. Exporters are advised to check with their importers to confirm the eligibility of particular establishments. Additional information regarding plant eligibility will be posted as it becomes available.
B. Halal certification must be issued by the registered Islamic organization that has been accredited by the Malaysian authorities. Routine follow-up inspections can be made by authorized Islamic authorities based in the U.S.

EXPORT REQUIREMENTS FOR MOROCCO

Eligible/Ineligible Products

A. Eligible products
   1. Red meat and meat products
   2. Poultry and poultry products. Meat and poultry exports to Morocco require an import license from the Ministry of Commerce. Importers obtain the import license following approval by the Ministry of Agriculture. U.S. exporters interested in sending products to Morocco should contact their importer for a copy of the import permit.

Documentation Requirements

A. For red meat products, FSIS Form 9060-5 – Meat and Poultry Export Certificate of Wholesomeness — should be obtained.
B. For poultry products, FSIS Form 9060-5 and FSIS Form 9352-1 — veterinary certificate for poultry meat exported into Morocco — should be obtained.
C. Signature on certificates — all required forms and supplementary statements must be dated and signed by an FSIS veterinarian [name, degree (DVM or equivalent) must be typed after the signature].
D. Ritual slaughter — a certificate of Islamic (Halal) slaughter must accompany shipments labeled "halal." Animals must be slaughtered in accordance with Islamic law.

Plants Eligible to Export

All federally inspected meat plants are eligible to export to Morocco.

EXPORT REQUIREMENTS FOR OMAN

Eligible/Ineligible Products

A. Eligible products
1. Fresh/frozen red meat and red meat products
2. Fresh/frozen poultry and poultry products

Note: Pork is eligible if clearly labeled.

Slaughter Requirements

Ritual Islamic halal slaughter requirements apply.

Labeling Requirements

A. Fresh/frozen meat and poultry — in addition to the labeling features mandatory in the U.S., packaged meat and poultry must bear the following features:
1. Production (slaughter or freezing) and expiration dates — date format requirements for Oman must conform to the following: day/month/year for products with a shelf life of 6 months or less and month/year for products with a shelf life of more than

6 months. Dating should be in numeric format and bilingual (English/Arabic).

2. Metric net weight — at present, there are no restrictions regarding net weight tolerances.
3. Country of origin.
4. Statement that the product has been slaughtered according to Islamic principles.
5. Bilingual labels — Arabic and English.
6. Product name — product names may deviate from U.S. standards if approved according to 9 CFR 317.7 or 381.128. For example, chicken from the U.S. is commonly labeled as "American Chicken."

B. Oman permits entry of pork products, but all pork products, including lard, must be identified on label.

Documentation Requirements

Certification requirements are as follows:

A. FSIS Form 9060-5 — Meat and Poultry Certificate of Wholesomeness — should be obtained. All FSIS Form 9060-5 certificates must be dated and have the signature and title of an FSIS veterinarian.

B. Ritual slaughter — in addition to FSIS certification, the exporter must obtain a certificate of Islamic halal slaughter from a member of an Islamic center or Islamic organization. A certificate of Islamic slaughter is a certificate issued by a member of a Muslim organization recognized by the importing country to provide this service. The certificate states that animals were slaughtered according to Muslim religious requirements. This certificate must accompany products labeled "halal." The certificate must be endorsed by the Arabian-American Chamber of Commerce or by Arabian Consulate and must accompany all shipments.

C. Additional Certification
1. Fresh/frozen unprocessed product — products bearing Halal label claims must be accompanied by an appropriate halal certificate or a written assurance from the exporter that an appropriate halal certificate will be supplied to accompany that shipment before it reaches its destination.
2. Raw materials used in processed products with halal label claims must be accompanied by an appropriate halal certificate.

Handling/Storage Requirements

Oman requires that instructions for consumers concerning storage, preparation, and other special handling requirements accompany all shipments.

Other Requirements

    A. Consignee — product must be consigned directly to Oman.

    B. Expiration period

        1. For frozen beef there is no fixed expiration time. Twelve months is suggested as a reasonable time frame for which to set an expiration date.

        2. For frozen poultry the expiration date should be no longer than 12 months.

Plants Eligible to Export

All federally inspected establishments are eligible to export to Oman. If products are to be labeled "halal," the plant must be able to accommodate the Islamic requirements.

EXPORT REQUIREMENTS FOR PAKISTAN

Eligible/Ineligible Products

Neither edible nor inedible products for export may contain products or by-products of pig, hog, boar, or swine.

Labeling Requirements

Bulk as well as the consumer packs of processed food items should be inscribed with the date of manufacture and the date of expiration.

Documentation Requirements

    A. Ritual slaughter — certificates must contain the statement: "The poultry covered by this certificate was slaughtered by means of a sharp knife cutting through the skin, jugular vein, and trachea to result in thorough bleeding out of the carcass in preparation for dressing and evisceration."

    B. Special statements — both edible and inedible products for export must conform to the following requirements:

1. The letter of credit must incorporate the condition that export items should not contain products or by-products of pig, hog, boar, or swine.
2. At the time of clearance, the exporter should submit before custom authorities a certificate from the supplier that the imported consignment does not contain products or by-products of pig, hog, boar, or swine.

EXPORT REQUIREMENTS FOR QATAR

Eligible/Ineligible Products

A. Eligible products
   1. Fresh/frozen red meat and meat products (excluding pork)
   2. Fresh/frozen poultry and poultry products
B. Ineligible products
   1. Pork and pork products

Labeling Requirements

For fresh/frozen meat and poultry, in addition to the labeling features mandatory in the U.S., packaged meat and poultry must bear the following features:

A. Production (slaughter) and expiration dates — date format requirements for Qatar must conform to the following: day/month/year for products with a shelf life of 6 months or less and month/year for products with a shelf life of more than 6 months. Dating should be in numeric format and bilingual (English/Arabic).
B. Arabic stickers are allowed.
C. A statement that the product has been slaughtered according to Islamic principles is required.
D. Metric net weight — presently, Qatar has a tolerance of 10% variation in the labeled net weight.

Documentation Requirements

Certification requirements are as follows:

A. FSIS Form 9060-5 — Meat and Poultry Certificate of Wholesomeness — should be obtained. All FSIS Form 9060-5 certificates must be dated and have the signature and title of an FSIS veterinarian.

B. Ritual slaughter (Islamic slaughter certification) — a certificate of Islamic (halal) slaughter is required. Exporters may obtain a certificate of Islamic slaughter from a member of an Islamic center or Islamic organization. A certificate of Islamic slaughter is a certificate issued by a member of a Muslim organization recognized by the importing country to provide this service. The certificate states that animals were slaughtered according to Muslim religious requirements. This certificate must accompany products labeled "halal." The certificate must be endorsed by the Arabian-American Chamber of Commerce and by the Qatar Consul and must accompany all shipments. The telephone numbers of the Arabian-American Chambers of Commerce are listed below.

C. Additional certification
1. Fresh/frozen unprocessed products bearing halal label claims must be accompanied by an appropriate halal certificate or a written assurance from the exporter that an appropriate halal certificate will be supplied to accompany that shipment before it reaches its destination.
2. For processed products with halal label claims, raw materials used in processed products with halal label claims must be accompanied by an appropriate halal certificate.

Handling/Storage Requirements

A. Special handling — Qatar requires that instructions for consumers concerning storage, preparation, and other special handling requirements accompany shipments.
B. Packaging — vacuum packaging is not required.

Other Requirements

A. Product arrival and shelf life — meat and poultry products must arrive within four months of slaughter date.
B. Qatar laboratory sampling — random samples are routinely collected from meat and poultry products entering Qatar. Product is examined for:
1. Pesticides
2. Salmonellae and other pathogenic bacteria
3. Total bacteria counts
4. Heavy metals
5. Species identification tests for pork tissue, including lard, in formulated product

C. Shelf life for frozen beef and poultry products is twelve months, for minced meats is nine months, for fresh beef is 21 days, and for fresh mutton is 14 days.

Plants Eligible to Export

All federally inspected establishments are eligible for export to Qatar.

EXPORTS REQUIREMENTS TO SAUDI ARABIA

Eligible/Ineligible Products

A. Eligible products
   1. Fresh/frozen red meat and further processed red meat products.
      a. Male cattle not over 5 years old must be in cuts not smaller than quarters. Sheep not over 3 years old must be shipped in whole carcasses.
      b. Each carcass (side or quarter if cattle) must:
         i.   Bear a legible U.S. inspection legend.
         ii. Be free from any preservatives.
         iii. Have kidneys removed.
         iv. Be wrapped in clean white cloth.
      c. The carcass shall be eviscerated and free from head, feet, and kidney fat. A part of the tail may be left to identify the animal type.
      d. No preservatives, antibiotics or coloring material, except the stamping ink, should be used.
   2. Fresh/frozen poultry and further processed poultry products. *Note:* Poultry meat and further processed poultry products must be derived from birds that have not been fed animal protein, animal fats, or animal by-products. To be eligible for export to Saudi Arabia, poultry must be produced under an Animal Protein-Free Certification Program for Poultry administered by the Agricultural Marketing Service (AMS), USDA.
B. Ineligible products
   1. Offal products, e.g., liver, tongue, brain, kidney, and stomach, from all species.

Labeling Requirements

A. All products — storage temperature must be given with the refrigeration statement on the boxes to fully clarify the type of product being handled (e.g., "Keep frozen — store at or below __°C" or "Keep chilled (or refrigerate) — store between __°C and __°C.")

B. Fresh/frozen meat and poultry — in addition to the labeling features mandatory in the U.S., precut/processed and packaged meat and poultry must bear the following features (in print):

1. Bilingual labels — the Arabic language must be one of the languages used for declaration.

2. A statement must be included on the label that the halal product has been slaughtered according to Islamic principles.

3. Metric net weight

   a. Poultry — chilled whole bird weights can range from 550 to 1800 g. Weight classes (grades) should be divided by a difference of at least 100 g between the classes. There is a tolerance of 50 g within a class.

   b. Frozen poultry — individual birds (units) within each box or carton should be uniform in size and weight, with a tolerance not exceeding 10%.

4. For production (slaughtering or freezing) and expiration dates, the name of the month should be spelt out or abbreviated (e.g., Jan or January 1985). Dates may also be expressed 27/4/87 or 27 Apr 87 in English and Arabic. Calendar strips may be pre-printed on the label, allowing the designation of calendar dates with the literal translation:

   a. Production (slaughtering or freezing) date must be accompanied by the statement "Production good for __ months from date of production."

   b. The expiration date for frozen poultry is calculated from the date the product was first frozen. Expiration dates for fresh/frozen beef, veal, buffalo, mutton, goat, and fresh poultry are calculated from the slaughter date.

5. The use of the terminology "Keep Refrigerated" is not acceptable on labels for frozen product.

6. The following labeling methods may be used as alternatives to comply with labeling requirements (Section B):

   a. Stickers — must not interfere with label terminology and be self-destructive on removal. Overlabeling may result in refused entry of product. Stick-on labels covering existing labeling information are in violation.

   b. Inserts — must be accompanied by production and expiration dates.

   c. Ink stamps — ink must be indelible and legible. (Ink stamps are the least desirable labeling method.)

C. Processed meat and poultry

1. Features required on processed meat and poultry labels:
   a. Bilingual labels with labeling features mandatory in the U.S.
   b. Metric net weights.
   c. Identification of pork products (including lard).
   d. Production and expiration dates (see "Other Requirements," Section C — "Product Arrival and Expiration Date").
2. A certificate of Islamic slaughter is not required for processed meat and poultry products. If processed products are labeled with halal claims, then origin raw materials must be from Muslim-approved operations and be accompanied with appropriate certificates supporting Islamic slaughter.

D. Prepackaged processed meat and poultry product requirements
   1. Production date (packaging or freezing) and expiration date (see "Other Requirements," Section C — "Product Arrival and Expiration Date").
   2. Net weight of frozen product.

Documentation Requirements

Certification requirements are as follows:

A. Beef and mutton
   1. FSIS Form 9060-5 — Meat and Poultry Certificate of Wholesomeness — should be obtained.
   2. The following statements must be provided as an FSIS letterhead certificate:
      a. "The meat is from animals of an average age of __ and is subject to veterinary examination not more than 12 hours prior to and immediately after slaughter and found free of disease and suitable for human consumption."
      b. "The animals originated from herds under state supervision for the diseases regulated by the state or the United States Department of Agriculture."
      c. "The animals for slaughter were given careful veterinary inspection and found free from clinical signs of infectious and contagious disease."
      d. "The United States has been free of foot-and-mouth disease (FMD) since 1929 and is free of bovine spongiform encephalopathy (BSE)."
      e. "The product was prepared, handled, stored, and transported under hygienic conditions."
      f. "The product is in conformity with pertinent United States regulations concerning production, storage, and transport."

g. "There have been no outbreaks of vesicular stomatitis, rinderpest, peste des petite ruminants, contagious bovine pleuropneumonia, lumpy skin disease, Rift Valley Fever, bluetongue, sheep pox, goat pox, Theileriosis hemorrhagic septicemia (Asian-type), bovine brucellosis, bovine tuberculosis, bovine babesiosis, contagious agalactia, and contagious caprine pleuropneumonia in the state of origin in the 3 months prior to slaughter."

h. "The product has not been stored with products that have not passed inspection."

i. "The feeding of ruminant protein (meat meal and bone meal) to ruminants is prohibited in the United States."

j. "Based on a national residue program, the product does not contain harmful residues of substances with hormonal effect."

Note: The meat letterhead certification statements have been discussed with Saudi import officials and are believed to meet Saudi certification requirements.

B. Poultry and poultry products

1. FSIS Form 9060-5 — Meat and Poultry Certificate of Wholesomeness — should be obtained.

2. The following statements must be provided as an FSIS letterhead certificate:

a. "The birds were subject to veterinary examination not more than 12 hours prior to and immediately after slaughter and found free from disease and suitable for human consumption."

b. "The birds originated from flocks under state supervision for the diseases regulated by the state or the United States Department of Agriculture."

c. "The birds for slaughter were given careful veterinary inspection and found free from clinical signs of infectious and contagious disease."

d. "The product was prepared, handled, stored and transported under hygienic conditions."

e. "The product is in conformity with pertinent United States regulations concerning production, storage, and transport."

f. "There have been no outbreaks of highly pathogenic avian influenza (fowl plaque), lethal-type Newcastle disease, fowl typhoid, and pullorum disease in the state of origin in the 3 months prior to slaughter."

g. "The product has not been stored with products that have not passed inspection."

h. "The poultry has not been fed with animal protein, animal fats, or animal by-products."

i. "The United States Food and Drug Administration prohibits the use of growth-stimulating hormones in poultry feeds."

Note: The poultry letterhead certification statements have been discussed with Saudi import officials and are believed to meet Saudi certification requirements.

C. Islamic slaughter certification — in addition to FSIS certification, the exporter must obtain a certificate of Islamic (halal) slaughter from a member of an Islamic center or Islamic organization. Products bearing halal label claims must be accompanied by an appropriate halal certification or a written assurance from the exporter that an appropriate Halal certificate will be supplied to accompany that shipment before it reaches its destination. However, FSIS is not responsible for certifying that products intended for export to Muslim countries meets appropriate requirements for religious slaughter. A certificate of Islamic slaughter is a certificate issued by a member of a Muslim organization recognized by the importing country to provide this service. The certificate states that animals were slaughtered according to Muslim religious requirements. Procedures for export certification do not include agency oversight of the halal process or review of the authenticity of the halal certificate. Processed products with halal claims should also be accompanied by an appropriate halal certificate. The certificate must be endorsed by the Arabian-American Chamber of Commerce or by Arabian Consul and must accompany all shipments.

D. Frozen product — the following statement must be given in the "Remarks" section of FSIS Form 9060-5: "Product was frozen 72 hours after slaughter."

E. Shipment for U.S. personnel — the certificate of Islamic slaughter may be waived if products are shipped for consumption by U.S. personnel in Saudi Arabia. Such shipments require a written statement (filed with export certificate) that the shipment is so destined, and full responsibility is accepted by the exporter for possible problems in gaining entry of the shipment into Saudi Arabia as certified.

F. All certificates must be dated and have the signature and title of an FSIS veterinarian.

Handling/Storage Requirements

Saudi Arabia requires that instructions for consumers concerning storage, preparation, and other special handling requirements accompany all shipments.

Other Requirements

A. Saudi Arabian import inspection:
1. Laboratory sampling — random samples collected on all meat and poultry products entering Saudi Arabia are examined for:
a. *Salmonellae* — product rejected when more than two of five subsamples are positive.
b. *E. coli* — no tolerance in ground beef.
c. Growth bacteria — maximum 10,000,000/g.
d. Volatile nitrogen — beef: maximum 20 mg/100 g; poultry: maximum 50 g/100 kg.
2. Species identification tests for pork are routinely run on all products.
3. When frozen poultry sample is thawed, the amount of water collected should not exceed 5%. Saudi Arabian officials recommend that U.S. industry run test prior to shipment to ensure that the product is not detained on arrival.
B. Detained product — if the product is detained, an appeal must be made in person by a Saudi Arabian broker or consignee to the Saudi Ministry of Commerce. Appeals are decided on a case-by-case basis.
C. Product arrival and expiration date
1. For frozen poultry, the period elapsed from slaughtering until arrival in Saudi Arabia should not be more than 3 months for frozen turkey, duck, goose, and chicken. Frozen poultry should be maintained at a temperature not more than −18°C, with an expiration date of 12 months from the slaughtering date.
2. For frozen red meats, the period elapsed from slaughtering and freezing until arrival in Saudi Arabia should not be more than 4 months. Products should be maintained frozen at a temperature not more than −18°C, with an expiration date of 12 months from slaughter date for beef, buffalo, mutton, and goat, and 9 months from freezing date for minced meat, hamburger, sausages, and livers.
3. For chilled red meats, the period elapsed from slaughtering until arrival to Saudi Arabia should not be more than 10 days at a

temperature not more than −2°C, with an expiration date of 4 weeks after slaughtering date.

4. For chilled poultry, the period elapsed from slaughtering until arrived to Saudi Arabia shall not be more than 7 days at a room temperature of −2°C.

5. For chilled vacuum meats and poultry, the period elapsed from slaughter until arrival in Saudi Arabia should not be more than 40 days at a temperature not more than −2°C, with an expiration date of 10 weeks after slaughtering date.

6. For canned meats, the expiration date is 24 months for meat and poultry products in sterilized, metallic containers.

7. Samples of meat and poultry products, regardless of the quantity, should comply with all labeling and other requirements. If samples are hand-carried, these requirements are often waived by the authorities on entry.

D. Product transiting third-country ports must be sealed with a USDA seal and the seal number and container number must be typed in the "Remarks" section of FSIS Form 9060-5.

Plants Eligible to Export

All federally inspected plants are eligible to export to Saudi Arabia.

EXPORT REQUIREMENTS FOR SINGAPORE

Eligible/Ineligible Products

A. Eligible products
1. Meat products
   a. Fresh/frozen beef
   b. Fresh (chilled)/frozen pork
   c. Fresh/frozen lamb
2. Poultry products
   a. Federally inspected fresh/frozen poultry
   b. Federally inspected frozen ducks may also be exported to Singapore under the following conditions:
      i.   With or without head and feet attached, trachea and esophagus attached, and oil glands intact.
      ii. Antemortem and postmortem inspections are performed, and product is prepared as ready-to-cook (except for head and feet attached and with or without trachea, esophagus, and oil glands).

   iii. Heads are completely defeathered and mouth and nasal passages thoroughly washed.

   iv. Water used for chilling poultry with trachea and esophagus attached is not used for chilling poultry with trachea and esophagus removed.

  c. Poultry feet (meeting Hong Kong standards)

  d. Ostrich meat, provided the birds were raised in the U.S. and did not come from indigenous and/or protected populations

 3. Processed products (including canned products)

B. Ineligible products

 1. Beef lungs

 2. Products imported into the U.S. from third countries

Labeling Requirements

A. Poultry feet — shipping containers must:

 1. Bear the wording "Chicken feet (turkey feet) for export to Singapore only, packed under sanitary supervision of USDA."

 2. Show establishment number, name, and address of plant and "USA." The official inspection legend will not be used on shipping containers.

B. For ducks (export of ducks with feet and heads attached), all labeling should fully describe the product and bear the wording "For export to Singapore only."

C. Shipping cartons of all fresh/frozen meat and poultry products must bear slaughter or production dates (month and year format, code dating is not acceptable).

D. All product names listed on FSIS Form 9060-5 must include species. If not normally indicated in the label name, such as "Prosciutto," it must be added in parenthesis after the name, e.g., "Prosciutto (pork)."

Processing Requirements

Poultry feet may be exported to Singapore under the following conditions (Hong Kong standard):

A. Feet must be removed after dressed poultry receives final wash, before entering the evisceration room or immediately after transfer from picking to the eviscerating conveyor line.

B. Feet must be handled sanitarily, packed in clean containers, and frozen promptly. Plant management should cooperate in proper handling of this product and instruct plant employees to reject any feet obviously unfit for food.

C. Feet must be scaled and toenails removed.

D. Plants in which feet are removed in the picking room must ensure that feet are not contaminated during collection.

E. The hock joint areas must be observed for swellings and abnormalities that might affect product wholesomeness or product packaging operation.

F. Feet should be prepared and packed in shipping cartons in a separate room under sanitary conditions and refrigerated until frozen.

Documentation Requirements

Certification requirements are as follows:

A. FSIS Form 9060-5 (05/06/1999) — Meat and Poultry Certificate of Wholesomeness— should be obtained.
   1. All frozen meat and poultry and meat and poultry products exported to Singapore must include the following statement in the "Remarks" section: "The product was frozen under USDA supervision." Acceptable methods of verifying the above statement are:
      a. Product was frozen at the exporting establishment.
      b. Product was frozen "off premises" and an "off-premises freezing quality control certificate for product shipped to Singapore" certificate is provided to FSIS as industry certification of freezing procedures. This form does not accompany the product to Singapore.

B. FSIS Form 9435-1 (9/93) should be obtained. When completing the FSIS Form 9435-1 (9/93), it should be ensured that the following information is correctly reflected on the certificate:
   1. Production dates on the certificate agree with production dates indicated on the product.
   2. Date of manufacture is prior to date of certification.
   3. Establishment number on the certificate agrees with establishment number indicated on the product.
   4. The export certificate number on the certificate agrees with the export certificate number stamped on the carton(s); in addition, the following requirements must be met:
      a. Correct manufacture and slaughter dates (with the month spelled out) must be listed on the FSIS Form 9435-1 (9/93), where requested.

i. Frozen beef, mutton, and poultry must arrive in Singapore within 6 months of slaughter. Product arriving in Singapore dated 6 to 12 months from the time of slaughter will be subject to automatic detention and subject to tests before sale. Product arriving in Singapore dated later than 12 months from time of slaughter will not be allowed entry.

ii. Frozen pork must arrive in Singapore within 3 months of slaughter. Product arriving in Singapore dated 3 to 6 months from the time of slaughter will be subject to automatic detention and subject to tests before sale. Products arriving in Singapore dated later than 6 months from time of slaughter will not be allowed entry.

iii. Processed pork, beef, mutton and poultry must arrive in Singapore within 3 months of manufacture. Product arriving in Singapore dated 3 to 6 months from time of manufacture will be subject to automatic detention and testing before sale. Product arriving in Singapore dated later than 6 months from time of manufacture will not be allowed entry. Also, prior approval from the primary production department (PPD) is required for specialty products (processed pork, beef, mutton, and poultry), which reflect manufacturing dates beyond the permissible time frame.

b. FSIS Form 9060-5 (05/06/1999) and FSIS Form 9435-1 (9/93) and all supplementary statements must be dated and signed by the same veterinarian. DVM or an equivalent degree must be typed or printed after the signature.

5. Beef and beef product certification — the following additional statement must be typed in the "Remarks" section on FSIS Form 9060-5: "The United States has been free from bovine spongiform encephalopathy for six years prior to the date of slaughter of the animals and the date of export of the product to Singapore."

6. Fresh (chilled) pork — trichinae testing must be done prior to entering the following statement in the "Remarks" section of FSIS Form 9060-5 (05/06/1999): "Each carcass from which the meat was derived was tested for trichina with negative results." Laboratories conducting trichinae analysis must be certified under an Agricultural Marketing Service (AMS) laboratory certification program. *Note:* Trichinae testing requirements do not apply to pork bacon products.

7. Ostrich meat certification — the following additional statements must be typed in the "Remarks" section on FSIS Form 9060-5: "The birds were raised in the U.S. and did not come from indigenous and/or protected populations." The country of origin must be stated.

8. Poultry feet (meeting Hong Kong standard) — poultry feet processed in the manner outlined previously are not considered as edible product in the U.S., but are considered edible when prepared for export to Singapore.

   a. FSIS Form 9060-5 (05/06/1999) should be obtained. The check-off box statement found on FSIS Form 9060-5 "I certify that the poultry and poultry products specified above come from birds that were officially given an ante-mortem and post-mortem inspection and passed in accordance with applicable laws and regulations of the United States Department of Agriculture and are wholesome and fit for human consumption" is *not* applicable to poultry feet produced according to the Hong Kong standard and should *not* be checked off.

   b. When the requirements of the Hong Kong standard are met, the inspector may issue an export certificate including the following statement: "This certifies that the poultry feet specified above have been processed in compliance with the Regulations Governing the Inspection of Poultry and Poultry Products (9 CFR Part 381) as promulgated by the Secretary of Agriculture, and are sound and unadulterated so far as can be determined by external examination and acceptable for human consumption according to Hong Kong standards." This certification may be typed in or immediately above the "Remarks" space on FSIS Form 9060-5 (05/06/1999).

   c. The certificate must be made by an inspector at plant of origin only.

   d. The certificate must bear the inspector's initials immediately after the certification. All FSIS Form 9060-5 certificates must be dated and have the signature and title of an FSIS veterinarian.

9. Processed products

   a. Canned meat and poultry products — the following additional statement must be typed on FSIS Form 9060-5 (05/06/1999) for canned meat and poultry products: "The canned meat (poultry) products described herein were manufactured and inspected in accordance with section 318.300 through 318.311 of the USDA regulations and have been subjected to a temperature of not less than __°C for a period of not less than __ minutes. This sterilizing process would have a sterilizing value (Fo) of __ minutes."

b. Canned pork and beans which are not amenable to the Meat Inspection Act may be certified under Part 350 of the regulations (certification service). The product must be accompanied by a declaration from the manufacturer stating:

i. The meat content of the product (including fat).

ii. That the product has been prepared from sound and wholesome ingredients.

iii. That the product has been heated to __°C for __ min.

iv. That every portion of the contents has been heated to a temperature of not less than 100°C.

c. The above declaration must be countersigned by an MPI veterinarian stating that he/she has no reason to doubt the truth of the manufacturer's declaration and that he/she is satisfied with the cleanliness and manufacturing practices of the processing plant. This certificate may be typed on a company letterhead. The certificate should include "MPI Veterinarian" under the veterinarian's signature. FSIS Form 9060-5 should not be issued.

10. Beef jerky certification — U.S. beef jerky sold in Singapore as a carry-on item to Japan must include the following statement in the "Remarks" section of FSIS Form 9060-5: "Oven temperature has been maintained at 190°F for the last thirty minutes or longer and this process insures the beef jerky meets the requirement of 70°C internal temperature."

11. Singapore requires importers of meat and poultry products to be officially registered. This is presented for information for U.S. exporters.

Other Requirements

A. Singapore performs microbiological testing on raw, fresh/frozen, and cooked products.

B. Weights — when product originates from two or more establishments, the weights and numbers of cartons must be divided to accurately reflect the amount of product originating from each establishment.

C. Fresh (chilled) pork

1. The product must be derived from barrows or gilts produced from rations containing no garbage or swill, and slaughtered without the application of carcass sprays. Most U.S. commercial

swine production and slaughter operations do not employ these practices, and export certification may generally be provided unless there are known instances of their use.

2. Fresh (chilled) pork must have a shelf life of at least 6 weeks.

D. Fresh/frozen pork must be treated to destroy trichinae by one of the methods in Section 318.10. *Note:* Trichinae treatment requirements do not apply to pork bacon products.

E. Reinspection of ready-to-cook duck carcasses with head and feet — defect descriptions, criteria, and procedures described in MPI Directive 9180.1 should be used with the sampling plan and its limits as shown in the following table:

Sample Size	Acceptable Numbers (Maximum)	Major Total
10[a]	3	30
Absolute limit per subgroup	4	34
Tightened criteria	2	27

[a] Cumulative sampling not required; random 10-bird samples used.

F. Export shipments to Singapore transiting foreign ports

1. Meat and poultry shipments from the U.S. to Singapore may be transshipped through a foreign port, providing the following conditions are met:

a. The consignment must be shipped in a refrigerated van container (reefer) sealed with an official USDA seal.

b. Containers may be sealed at the origin plant or at ports, i.e., in the container staging yards or on the container ships.

c. The serial number of the seal must be recorded on the export certificate or on the modified FSIS Form 7350-1, whichever the exporter requests.

d. The inspection officials in the transit port must certify the following statement on the original export certificate or on the modified FSIS Form 7350-1: "Container has remained under refrigeration in the container yard within the port area of (city) throughout its stay in (city)."

e. The consignment must not stay in the transit port for more than 14 days.

f. The temperature of the container must be recorded throughout its journey from the U.S. to Singapore, and the temperature record chart must be provided for examination in Singapore.

    g. The steamship lines should:
- i. Contact inspection officials in Singapore to obtain agreement on proposed transit ports and procedures. [Transit ports currently approved are Hong Kong, Japan, and Keelung (Taiwan)].
- ii. Arrange with transit port inspection authorities for transit shipment supervision and certification.
- iii. Notify MPI personnel, in advance, when containers will be ready for sealing at the U.S. port area.
- iv. Provide seamen to assist the MPI inspector in sealing the containers on the container ship.
- v. Forward (or the exporter forwards) the original export certificate or the FSIS Form 7350-1, as applicable, to the transit port for certification, and subsequently to the importer in Singapore.

    h. The exporter must obtain the required documentation, arrange for the sealing of the containers with local inspection personnel (when the procedure has not previously been established), and fulfil the conditions which apply to transit shipments.

    i. Inspection officials will record seal number(s) on the original export certificate or on the FSIS Form 7350-1, as applicable.

Plants Eligible to Export

All federally inspected establishments are eligible to export to Singapore.

EXPORT REQUIREMENTS FOR TURKEY

Eligible/Ineligible Product

A. Eligible products
1. Canned meat products
B. Ineligible products
1. Fresh/frozen red meat/poultry and processed red meat/poultry are not eligible at this time. The animal disease statements required by the Turkish government cannot be certified to by FSIS. The Export Coordination Division should be contacted for additional information.

Documentation Requirements

Certification requirements for canned meat products — obtain FSIS Form 9060-5, Meat and Poultry Export Certificate of Wholesomeness.

Other Requirements

A. Canned meat product(s) — the following two additional documents are required:
   1. Certificate of origin, indicating where the food originated.
   2. Certificate of analysis, which includes a physical analysis, chemical analysis, and microbiological analysis. This analysis not only indicates meat and products in the food but also identifies any additives (with their identification or E-numbers) and quantities in the product. The values with regard to the pathogen microorganisms should also be included in the analysis.
B. Ships' stores — meat and poultry products intended for ships' stores are eligible from any federally inspected facility. The product should be certified with FSIS Form 9060-5. The statement "Product intended for ships" stores should be typed in the "Remarks" section.

EXPORT REQUIREMENTS FOR THE UNITED ARAB EMIRATES (UAE)

Eligible/Ineligible Products

A. Eligible products
   1. Fresh/frozen red meat and poultry
   2. Red meat and poultry products

Slaughter Requirements

Ritual Islamic halal slaughter requirements apply.

Labeling Requirements

A. All products — storage temperature must be given with the refrigeration statement on the label to fully clarify the type of product being handled. (e.g., "Keep frozen — store at or below __°C" or "Keep chilled (or refrigerated) — store between __°C and __°C.")
B. Fresh/frozen meat and poultry — in addition to the labeling features mandatory in the U.S., precut and packaged meat and poultry must bear the following features (in print):
   1. Production (slaughtering or freezing) and expiration dates.
      a. Date format requirements for UAE must conform to the following: day/month/year for products with a shelf life of 6 months or less and month/year for products with a shelf life of more than 6 months. Dating should be in numeric format.

b. The expiration date must be calculated from the date the product was frozen. *Exception:* The UAE permits expiration dates on bagged poultry to be printed on adhesive tape wrapped around metal clip area.

2. Statement that the product has been slaughtered according to Islamic principles. *Exception:* The UAE does not require reference to Islamic slaughter on consumer packages, but exporters should be aware that such products will have limited distribution.

3. Shelf life of product — shelf-life limits have been placed on chilled vacuum-packaged meat, frozen meat, and other meat/poultry products. Fast-spoiling foods with a shelf life not exceeding 3 months must have complete date stated on the label. Use of the terminology "Better to use before..." on label is not accepted.

4. Country of origin.

5. Metric net weight labeling is required. At present, there are no restrictions regarding net weight tolerances.

6. Alcoholic materials and species of animal fats, gelatin, food additives, and blood must be declared on label when the product contains such materials.

C. The following methods of labeling are alternatives to meeting the requirements for labeling packaged fresh/frozen meat and poultry:

1. Sticker may be used but must not obliterate label terminology and be self-destructive on removal. Stick-on labels covering required label features are not permitted.

2. Inserts, if used, must be accompanied by production and expiration dates. Inserts must be made of approved materials.

D. Canned goods — expiration and production dates must be pre-printed on the labels.

Documentation Requirements

Certification requirements are as follows:

A. FSIS Form 9060-5 should be obtained. All FSIS Form 9060-5 certificates must be dated and have the signature and title of an FSIS veterinarian.

B. Ritual slaughter (Islamic slaughter certification) — in addition to FSIS certification, the exporter must obtain a certificate of Islamic (halal) slaughter from a member of an Islamic center or Islamic organization. A certificate of Islamic slaughter is a certificate issued by a member of a Muslim organization recognized by the importing country to provide this service. The certificate states that animals were slaughtered according to Muslim religious requirements. The certificate must accompany products labeled "halal." The certificate must be endorsed by the Arabian-American Chamber of Commerce or by the Arabian Consulate and must accompany all shipments.

C. Additional certification

1. On fresh/frozen unprocessed product, halal label claims must be accompanied by an appropriate halal certificate or a written assurance from the exporter that an appropriate halal certificate will be supplied to accompany that shipment before it reaches its destination.

2. On processed products with halal label claims, raw materials used must be accompanied by an appropriate halal certificate.

Handling/Storage Requirements

A. The UAE requires that instructions for consumers concerning storage, preparation, and other special handling requirements accompany all shipments. These instructions should be addressed to the UAE municipality.

B. Poultry must be packaged in clear plastic packaging materials.

Other Requirements

A. The product must arrive in the UAE at least 3 months before the expiration date.

B. Expiration period

1. For frozen beef, the UAE has no fixed expiration time period. Twelve months is suggested as a reasonable period to set the expiration date.

2. For frozen poultry, the UAE has no fixed expiration time period. Nine months is suggested as a reasonable period to set the expiration date.

3. Chilled vacuum-packed meat/poultry has an expiration period of 3 months.

4. The shelf life (expiration period) for other meat/poultry products must not exceed 3 months.

Plants Eligible to Export

All federally inspected establishments are eligible to export to the UAE. If products are to be labeled "halal," the plant must be able to accommodate the Islamic requirements.

Appendix K

RECOMMENDED RITUAL SLAUGHTER GUIDELINES

Source: This is a draft document. For a revised version contact the American Meat Institute or www.grandin.com Based on a scientific review of information available on kosher and halal slaughter from Food Marketing Institute/National Council of Chain Restaurants. With permission. Prepared by Joe M. Regenstein, Professor, Food Science, and Head, Cornell Kosher Food Initiative, Department of Food Science, Cornell University, Ithaca, NY (jmr9@cornell.edu), with significant input from Dr. Temple Grandin, Associate Professor, Animal Sciences, Colorado State University, Fort Collins, CO.

INTRODUCTION

Kosher and halal slaughters are done to satisfy the requirements of the Jewish and Muslim communities, respectively. Much of the process is identical to that used for non-religious slaughter. Therefore, the focus of this document will be on the handling of the animal just prior to slaughter and during the actual slaughter process.

AMERICAN MEAT INSTITUTE DOCUMENTATION

The American Meat Institute (AMI) has had a recommended animal handling guide for meat packers (*AMI Meat Packer Guide*, 1991) for over 11 years. Dr. Temple Grandin has written these guidelines. The Animal Welfare Committee of the Food Marketing Institute/National Council of Chain Restaurants (FMI/NCCR) has approved the AMI documents, audit forms, and an audit system to implement these guidelines.

The current document assumes that regular audits associated with the AMI guidelines for the particular animal type being slaughtered are also taking place. These regular audits will deal with animal handling, including misuse of the electric prod, vocalization, and slipping or falling. The audit forms can be obtained from www.fmi.org or www.nccr.net.

AMERICAN MEAT INSTITUTE STANDARDS FOR KOSHER AND HALAL SLAUGHTER

The latest revision of this document, like the previous documents, deals with kosher (Jewish) and halal (Muslim) slaughter. This document has much to say about ritual slaughter. The complete text related to religious slaughter is included here along with suggested guidelines that can be understood from this document.

A second paper is reviewed in detail. This 1994 paper from *Meat Focus International* is titled "Religious slaughter and animal welfare: a discussion for meat scientists," for which the current author is the second author.

Dr. Grandin's web site (www.grandin.com) contains both documents. The site also contains a picture of a "pipe and rail" type of restraining device for small animals, and many drawings of various restrainer systems. A few are included in this document, but one is encouraged to go to her site for additional information.

The program staff of the Northeast Sheep and Goat Program (NESGP) at Cornell is currently working with Dr. Grandin on two different new designs for low-cost devices to handle small animals in an upright position based on the double rail, which we hope to post on her site as well as the NESGP and Cornell sites.

This document contains the guidelines established by the FMI/NCCR Animal Welfare Committee to implement the recommendations of the above two papers. The guidelines are presented in two ways: initially as suggested guidelines that come out of the two documents reviewed in detail so that the justification for these guidelines is clear, and the suggested guidelines are then compiled and edited into a comprehensive set of actual guidelines that appear as a separate section following the review of the two documents.

Recommended ritual slaughter practices (kosher and halal) — for both humane and safety reasons, plants which conduct ritual slaughter should install modern upright restraining equipment. The practice of hanging live cattle, calves, or sheep upside down should be eliminated. There are many different types of humane restraint devices available (*AMI Meat Packer Guide*, 1991).

Suggested guidelines — kosher and halal slaughter will be performed while the animal is upright. Handling systems that turn the live animal upside down and/or hang the animal are considered unacceptable.

Examples of acceptable systems include:

ASPCA pen — this device consists of a narrow stall with an opening in the front for the animal's head. After the animal enters the box, it is nudged forward with a pusher gate and a belly lift comes up under the brisket. The head is restrained by a chin lift for the rabbi or Muslim slaughter man prior to performing shehita or Muslim slaughter, respectively. Vertical travel of the belly life should be restricted to 28 inches so that it does not lift the animal off the floor. The rear pusher gate should be equipped with either a separate pressure regulator or special pilot-operated check values to allow the operator to control the amount of pressure exerted on the animal. The pen should be operated from the rear toward the front. Restraining of the head is the last step. The operator should avoid sudden jerking of the controls. Many cattle will stand still if the box is slowly closed up around them and less pressure will be required to hold them. Ritual slaughter should be performed immediately after the head is restrained (*AMI Meat Packer Guide,* 1991).

Suggested guidelines — those plants using an ASPCA pen should demonstrate that the system is operating properly, which includes demonstrating that the belly lift cannot lift the animal off the ground, that the rear pusher gate can be operated independently, that the animals are calm during the procedure, and that the slaughter takes place immediately after the head holder is in place. This guideline applies specifically to ASPCA-type pens. A more general guideline for all types of single animal restrainers appears at the end of this document.

An ASPCA pen can be easily installed in one weekend with minimum disruption of plant operations. It has a maximum capacity of 100 cattle per hour and it works best at 75 head per hour. A small version of this pen could be easily built for calf plants (*AMI Meat Packer Guide,* 1991).

Suggested guidelines — the ASPCA pen or other box-type restrainers that hold a single, large bovine shall be operated at less than 100 cattle per hour.

Conveyor restrainer system — either a V-restrainer or a center track conveyor restrainer can be used for holding cattle, sheep, or calves in an upright position during shehita or halal slaughter. Conveyor systems must completely support the animal's body in a comfortable upright position. The restrainer is stopped for each animal and a head holder holds the head for the ritual slaughter man. Research in Holland indicates that the center track design provides the advantage of reducing bloodspots in the meat.

For cattle, a head holder similar to the front of the ASPCA pen can be used on the center track conveyor restrainer. A biparting chin lift is attached to two horizontal sliding doors (*AMI Meat Packer Guide,* 1991).

Suggested guidelines — conveyor restrainer systems should be operated in a manner consistent with their design. Slaughter must take place immediately after the head holder is in place. More specific guidelines are given at the end of this document.

Small restrainer systems — for small locker plants which ritually slaughter a few calves or sheep per week, an inexpensive restrainer constructed from pipe can be used to hold the animal in a manner similar to the center track restrainer. Animals must be allowed to bleed out and become completely insensible before any other slaughter procedure is performed.

More specific guidelines are given at the end of this document.

Dr. Grandin maintains a web site at www.grandin.com. It contains a picture of a "pipe and rail" type of restraining device for small animals. The program staff of the Northeast Sheep and Goat Program at Cornell is currently working with Dr. Grandin on two different designs for low-cost devices to handle small animals in an upright position based on the double rail design.

Furthermore, the AMI has the following important safety reminder for religious slaughter. It speaks to the danger to people with respect to improper religious slaughter.

Recommended ritual slaughter practices: safety tips for works — shackling and hoisting large cattle and calves can be very dangerous. It has caused many serious accidents such as loss of an eye, permanent knee damage and head injuries from kicking and falling shackles. In one plant, replacement of the shackle hoist with a restrainer resulted in a 500 percent reduction in accidents. Shackling and hoisting of live sheep is also hazardous. There have been several incidents of teeth knocked out (*AMI Meat Packers Guide*, 1991).

Although the *AMI Meat Packers Guide* is a good starting point, a number of other issues arise that must also be considered.

FURTHER INFORMATION ABOUT RELIGIOUS SLAUGHTER

Drs. Grandin and Regenstein prepared a fairly comprehensive review of religious slaughter in a 1994 paper for *Meat Focus International*, titled "Religious slaughter and animal welfare: a discussion for meat scientists" (Grandin and Regenstein, 1994). It is available on Dr. Grandin's website, www.grandin.com. Large parts of this paper are relevant to the work of the FMI/NCCR Animal Welfare and the document, with editing, is presented below along with guidelines that suggest themselves from this work. In some cases the suggested guidelines are broken out separately, whereas in other cases they are left in the text as bolded materials. Italics are used for emphasis by the author and do not indicate guidelines.

Both the Muslim and Jewish faiths have specific requirements for the slaughter of religiously acceptable animals. The major difference from the general practices in most countries is that the animals are not stunned prior to slaughter.

Any Muslim may slaughter an animal while invoking the name of Allah. Again stunning prior to slaughter is generally not the practice. However, a nonpenetrating concussion stunning prior to slaughter has received approval from some Muslim authorities. Work in the 1980s in New Zealand led to the development of a very sophisticated electrical stunning apparatus that met a Muslim standard where an animal must be able to regain consciousness in less than a minute and must be able to eat within five minutes. Head-only electric stunning prior to Muslim slaughter is used in almost all sheep slaughter plants in New Zealand and Australia. Electric stunning of cattle is used in many New Zealand Muslim cattle slaughter plants and the practice is spreading to Australia. With Muslim slaughter in countries not using stunning, we are also concerned about the training given to the slaughter men. More work is needed on training programs to teach proper sharpening of knives and to improve the actual slaughter techniques (Grandin and Regenstein, 1994).

Suggested guidelines — the FMI/NCCR Animal Welfare Committee recommends that a knife, similar to the kosher chalef, be used for halal slaughter. The knife should have a straight blade and be at least twice the width of the neck. It should be kept sharp at all times.

A knife for halal slaughter of sheep and goats that is available from at least one knife supplier has been identified and preliminary work in actual slaughter plants has been successful. A special poster for on-farm humane (halal) slaughter has also been prepared.

In the case of the Jewish dietary laws, a specially trained person of known religiosity carries out the slaughter. This person, the "shochet," is specifically trained for this purpose. He is trained to use a special knife, called the "chalef," to rapidly cut in a single stroke the jugular vein and the carotid artery without burrowing, tearing, or ripping the animal. The knife is checked regularly for any imperfections that would invalidate the slaughter. This process when done properly leads to a rapid death of the animal. A sharp cut is also known to be less painful (Grandin and Regenstein, 1994).

NEED FOR OBJECTIVE EVALUATION

Given the importance of religious slaughter to people of these two major faiths, it is important that scientists be absolutely objective when evaluating these practices from an animal welfare standpoint.

Evaluation of religious slaughter is an area where many people have lost scientific objectivity. This has resulted in biased and selective reviewing of the literature. Politics has interfered with good science. There are three basic issues: stressfulness of restraint methods, pain perception during the incision, and latency of onset of complete insensibility.

Restraint

A key intellectual consideration is separation of the variable of restraint stress from the animal's reaction to the slaughter procedure. Stressful or painful methods of restraint mask the animal's reactions to the throat cut. In North America, some kosher slaughter plants use very stressful methods of restraint, such as shackling and hoisting fully conscious cattle by one rear leg.

Observations of the first author indicate that cattle restrained in this manner often struggle and bellow, and the rear leg is bruised. In Europe, the use of casting pens which invert cattle onto their backs completely mask reactions to the throat cut. Cattle resist inversion and twist their necks in an attempt to right their heads. Earlier versions of the Weinberg casting pen are more stressful than an upright restraint device (Dunn, 1992). An improved casting pen, called the Facomia pen, is probably less stressful than older Weinberg's pens, but a well-designed upright restraint system would be more comfortable for cattle. Another problem with all types of casting pens is that both cattle and calves will aspirate blood after the incision. This does not occur when the animal is held in an upright position.

Unfortunately some poorly designed upright American Society for the Prevention of Cruelty to Animals (ASPCA) restraint boxes apply excessive pressure to the thoracic and neck areas of cattle. In the interest of animal welfare the use of any stressful method of restraint should be eliminated. A properly designed and operated upright restraint system will cause minimum stress. Poorly designed systems can cause great stress. Many stress problems are also caused by rough handling and excessive use of electric prods. The very best mechanical systems will cause distress if operated by abusive, uncaring people.

In Europe there has been much concern about the stressfulness of restraint devices used for both conventional slaughter (where the bovine is stunned) and ritual slaughter. Ewbank et al. (1992) found that cattle restrained in a poorly designed head holder, i.e., where over 30 seconds was required to drive the animal into the holder, had higher cortisol levels than cattle stunned with their heads free. Cattle will voluntarily place their heads in a well-designed head restraint device that is properly operated by a trained operator (Grandin, 1992). Tume and Shaw (1992) reported very low cortisol levels of only 15 ng/ml in cattle during stunning and slaughter. Their measurements were made in cattle held in a head restraint (personal communication, Shaw, 1993). Cortisol levels during on-farm restraint of extensively reared cattle range from 25 to 63 ng/ml (Mitchell et al., 1988; Zavy et al., 1992).

For ritual slaughter or captive bolt stunning devices to restrain the body are strongly recommended. Animals remain calmer in head restraint devices when the body is also restrained. Stunning or slaughter must occur within 10 seconds after the head is restrained (Grandin and Regenstein, 1994).

Suggested guideline — the actual time between restraint of the head and actual slaughter must be less than 10 sec.

Reactions to the Throat Cut

The variable of reactions to the incision must be separated from the variable of the time required for the animal to become completely insensible. Recordings of EEG or evoked potentials measure the time required for the animal to lose consciousness. They are not measures of pain. Careful observations of the animal's behavioral reactions to the cut are one of the best ways to determine if cutting the throat without prior stunning is painful. The time required for the animals to become unconscious will be discussed later.

Observations of over 3000 cattle and formula-fed veal calves were made by the first author in three different U.S. kosher slaughter plants. The plants had state-of-the-art upright restraint systems. The systems are described in detail in Grandin (1988, 1991, 1992, 1993a, 1994). The cattle were held in either a modified ASPCA pen or a double rail (center track) conveyor restrainer.

This equipment was operated by the first author or a person under her direct supervision. Very little pressure was applied to the animals by the rear pusher gate in the ASPCA pen. Head holders were equipped with pressure limiting devices. The animals were handled gently and calmly. It is impossible to observe reactions to the incision in an agitated or excited animal. Blood on the equipment did not appear to upset the cattle. They voluntarily entered the box when the rear gate was opened. Some cattle licked the blood.

In all three restraint systems, the animals had little or no reaction to the throat cut. There was a slight flinch when the blade first touched the throat. This flinch was much less vigorous than an animal's reaction to an eartag punch. There was no further reaction as the cut proceeded. Both carotids were severed in all animals. Some animals in the modified ASPCA pen were held so loosely by the head holder and rear pusher gate that they could have easily pulled away from the knife.

These animals made no attempt to pull away. In all three slaughter plants, there was almost no visible reaction of the animal's body or legs during the throat cut. Body and leg movements can be easily observed in the double rail restrainer because it lacks a pusher gate and very little pressure is applied to the body. Body reactions during the throat cut were much fewer than the body reactions and squirming that occurred during testing of various chin lifts and forehead hold-down brackets. Testing of a new chin lift required deep, prolonged invasion of the animal's flight zone by a person. Penetration of the flight zone of an extensively raised animal by people will cause the animal to attempt to move away (Grandin, 1993a). The throat cut caused a much smaller reaction than penetration of the flight zone. It appears that the animal is not aware that its throat has been cut. Bager et al. (1992) reported a similar observation with calves. Further observations of 20 Holstein, Angus, and Charolais bulls indicated that they did not react to the cut. The bulls were held in a comfortable head restraint with all body restraints released. They stood still during the cut and did not resist head restraint. After the cut the chin lift was lowered, the animal either immediately collapsed or it looked around like a normal alert animal. Within 5 to 60 seconds, the animals went into a hypoxic spasm and sensibility appeared to be lost. Calm animals had almost no spasms and excited cattle had very vigorous spasms. Calm cattle collapsed more quickly and appeared to have a more rapid onset of insensibility. Munk et al. (1976) reported similar observations with respect to the onset of spasms. The spasms were similar to the hypoxic spasms [that] occur when cattle become unconscious in a V-shaped stanchion due to pressure on the lower neck. Observations in feedyards by the first author during handling for routine husbandry procedures indicated that pressure on the carotid arteries and surrounding areas of the neck can kill cattle within 30 seconds (Grandin and Regenstein, 1994).

Suggested guideline — animals shall become insensible in less than a minute.

The details spelled out in Jewish law concerning the design of the knife and the cutting method appear to be important in preventing the animal from reacting to the cut. The knife must be razor sharp and free of nicks. It is shaped like a straight razor and it must be twice the width of the animal's neck. The cut must be made without hesitation or delay. It is also prohibited for the incision to close back over the knife during the cut. This is called halagramah (digging) (Epstein, 1948). The prohibition against digging appears to be important in reducing the animal's reaction to the cut. Ritual slaughtermen must be trained in knife sharpening. Shochets have been observed using a dull knife. They carefully obeyed the religious requirements of having a smooth, nick-free knife, but they had failed to keep it sharp. Observations of halal cattle slaughter without stunning done by a Muslim slaughterman with a large, curved skinning knife resulted in multiple hacking cuts. Sometimes there was a vigorous reaction from the animal (Grandin and Regenstein, 1994).

Suggested guideline — a knife that is twice as long as the width of the neck and properly sharpened must always be used.

Further observations of kosher slaughter conducted in a poorly designed holder, i.e., one which allowed the incision to close back over the knife during the cut, resulted in vigorous reactions from the cattle during the cut. The animals kicked violently, twisted sideways, and shook the restraining device. Cattle which entered the poorly designed head holder in an already excited, agitated state had a more vigorous reaction to the throat cut than calm animals. These observations indicated that head holding devices must be designed so that the incision is held open during and immediately after the cut. Occasionally, a very wild, agitated animal went into a spasm that resembled an epileptic seizure immediately after the cut. This almost never occurred in calm cattle (Grandin and Regenstein, 1994).

Suggested guideline — the head holder must hold the incision open during and immediately after the cut, but excessive stretching of the neck that could cause tearing of the incision must be avoided.

Time to Loss of Consciousness

Scientific researchers agree that sheep lose consciousness within 2 to 15 seconds after both carotid arteries are cut (Nangeroni and Kennett, 1963; Gregory and Wotton, 1984; Blackmore, 1984). However, studies with cattle and calves indicate that most animals lose consciousness rapidly; however, some animals may have a period of prolonged sensibility (Blackwore, 1984; Daly et al., 1988) that lasts for over a minute. Other studies with bovines also indicate that the time required for them to become unconscious is more variable than for sheep and goats (Munk et al., 1976; Gregory and Wotten, 1984). The differences between cattle and sheep can be explained by differences in the anatomy of their blood vessels.

Observations by the first author, of both calf and cattle slaughter, indicate that problems with prolonged consciousness can be corrected. When a shochet uses a rapid cutting stroke, 95% of the calves collapse almost immediately (Grandin, 1987). When a slower, less decisive stroke was used, there was an increased incidence of prolonged sensibility. Approximately 30% of the calves cut with a slow knife stroke had a righting reflex and retained the ability to walk for up to 30 seconds.

Gregory (1988) provided a possible explanation for the delayed onset of unconsciousness. A slow knife stroke may be more likely to stretch the arteries and induce occlusion. Rapid loss of consciousness will occur more readily if the cut is made as close to the jaw bone as religious law will permit, and the head holder is loosened immediately after the cut. The chin lift should remain up. Excessive pressure applied to the chest by the rear pusher gate will slow bleed out. Gentle operation of the restrainer is essential. Observations indicate that calm cattle lose consciousness more rapidly and they are less likely to have contracted occluded blood vessels. Calm cattle will usually collapse within 10 to 15 seconds (Grandin and Regenstein, 1994).

Upright Restraint Equipment Design

Good upright restraint equipment is available for low stress, comfortable restraint of sheep, calves and cattle (Giger et al., 1977; Westervelt et al., 1976; Grandin, 1988, 1991, 1992, 1993). To maintain a high standard of animal welfare, the equipment must be operated by a trained operator who is closely supervised by plant management. Handlers in the lairage and race areas must handle animals gently and induce each animal to calmly enter the restrainer. Unfortunately, some very poorly designed restraint systems have recently been installed in Europe. The designers had little regard for animal comfort. In the following list of specific recommendations, the guideline points are bolded.

- All restraint devices should use the concept of optimal pressure. **The device must hold the animal firmly enough to provide a "feeling of restraint" but excessive pressure that would cause discomfort should be avoided.** Many people operating pens make the mistake of squeezing an animal harder if it struggles. Struggling is often a sign of excessive pressure.
- **To prevent excessive bending of the neck, the bovine's forehead should be parallel to the floor.** This positions the throat properly for ritual slaughter and stretches the neck skin minimizing discomfort. There is an optimal tightness for the neck skin. If it is too loose, cutting is more difficult. If it is too tight, the Jewish rule that prohibits tearing may be violated as the incision would have a tendency to tear before being cut by the knife. This also would be likely to cause pain. Some head restraints cause great distress to the cattle due to excessive bending of the neck in an attempt to obtain extreme throat skin tightness. This is not necessary for compliance with religious law. One must remember that 4000 years ago hydraulic devices which could achieve such extremes of throat tightness were not available. **All head holders must be equipped with pressure limiting devices.** Pressure limiting valves will automatically prevent a careless operator from applying excessive pressure. A 15 cm wide forehead bracket covered with rubber belting will distribute pressure

uniformly and the animal will be less likely to resist head restraint. The forehead bracket should also be equipped with an 8 cm diameter pipe that fits behind the poll. This device makes it possible to hold the head securely with very little pressure.

■ **The rear pusher gate of the ASPCA pen must be equipped with a pressure limiting device.** The animal must not be pushed too far forward in the head holder. The pressure must be regulated so that the animal stands on the floor with its back level. Arching of the back is a sign of excessive pressure. A calm relaxed animal will stand quietly in the pen and will not attempt to move its head. If the animal struggles, this is due to excessive pressure or being thrown off balance by the pusher gate.

■ **The animal must not be lifted off the floor by the belly lift of an ASPCA pen** because it does not fully support the body. Lifting devices that fully support the body in a comfortable upright position are permitted. In an ASPCA pen, the lift is for restraint not lifting. Lift travel should be restricted to 71 cm from the floor to the top of the lift. Other restrainers such as the double rail system are designed to give full support under the belly. The conveyor slats must be shaped to fit the contours of the animal's sternum in systems where an animal straddles a conveyor.

■ All parts of the equipment should always move with a slow steady motion. Jerky motions or sudden bumping of the animal with the apparatus excites and agitates them. **Jerky motion can be eliminated by installing flow control valves or other control devices.** These valves automatically provide a smooth steady motion even if the operator jerks the controls. **All restraint devices should use the concept of optimal (*not* maximum) pressure.** Sufficient pressure must be applied to give the animal a feeling of being held, but excessive pressure that causes struggling must be avoided. Animals will often stop struggling when excessive pressure is slowly reduced.

■ **All equipment must be engineered to reduce noise.** Air hissing and clanging metal noises cause visible agitation in cattle. Air exhausts must be muffled or piped outside. Plastic guides in the sliding doortracks will further reduce noise.

■ **A solid barrier should be installed around the animal's head to prevent it from seeing people and other distractions in its flight zone.** This is especially important for extensively reared cattle, particularly when they are not completely tame. On conveyor systems the barrier is often not required because the animals feel more secure because they are touching each other.

■ **Restraint equipment must be illuminated to encourage animals to enter.** Lighting mistakes or air blowing back at the animals will cause cattle to balk (Grandin, 1993b). **Distractions that cause balking must be eliminated.**

For plants [that] slaughter small numbers of sheep and goats a simple upright restrainer can be constructed from pipe (Giger et al., 1977). For veal calf plants a small ASPCA pen can be used. For large high speed plants a double rail restrainer can be equipped with a head holding device.

Some rabbinical authorities prefer inverted restraint and cutting downward because they are concerned that an upward cut may violate the Jewish rule that forbids excessive pressure on the knife. There is concern that the animal may tend to push downward on the knife during an upward cut. Observations indicate that just the opposite happens. When large 800 to 950 kg bulls are held in a pneumatically powered head restraint [that] they can easily move, the animals pull their heads upwards away from the knife during a miscut. This would reduce pressure on the blade. When the cut is done correctly, the bulls stood still and did not move the head restraint. Equal amounts of pressure were applied by the forehead bracket and the chin lift.

Upright restraint may provide the additional advantage of improved bleed out because the animal remains calmer and more relaxed. Observations indicate that a relaxed, calm animal has improved bleedout and a rapid onset of unconsciousness. Excited animals are more likely to have a slower bleedout. The use of a comfortable upright restraint device would be advantageous from a religious standpoint because rapid bleedout and maximum loss of blood obeys the biblical principle.

Rapid bleedout and a reduction in convulsions provide the added advantage of reducing petechial haemorrhages and improving safety. Convulsing animals are more likely to injure plant employees. A calm, quiet animal held in a comfortable restraint device will meet a higher animal welfare standard and will have a lower incidence of petechial haemorrhages.

Welfare Aspects of Slaughter

Many welfare concerns are centered on restraint. In Europe and the U.S. highly stressful restraint devices are still being used. Many of these systems apply excessive pressure or hold the animal in a position that causes distress. Proper design and operation of restraint devices can alleviate most of these concerns with cattle and sheep.

Restraint devices will perform poorly from an animal welfare standpoint if the animals balk and refuse to enter due to distractions such as shadows, air hissing or poor illumination Grandin, 1996. These easily correctable problems will ruin the performance of the best restraint system. Abusive workers will cause suffering in a well-designed system. For more information about properly operating pens, see Grandin, 1993.

Restraint devices are used for holding animals both for ritual slaughter and for conventional slaughter where animals are stunned. The use of a head restraint will improve the accuracy of captive bolt stunning. In large beef slaughter plants without head restraint captive bolt stunning has a failure rate of 3 to 5, i.e., a second shot is required.

Captive bolt and electric stunning will induce instantaneous insensibility when they are properly applied. However, improper application can result in significant stress. All stunning methods trigger a massive secretion of epinephrine (Van der Wal, 1978; Warrington, 1974). This outpouring of epinephrine is greater than the secretion that would be triggered by an environmental stressor or a restraint method. Since the animal is expected to be unconscious, it does not feel the stress. One can definitely conclude that improperly applied stunning methods would be much more stressful than kosher slaughter with the long straight razor sharp knife. Kilgour (1978), one of the pioneers in animal welfare research, came to a similar conclusion on stunning and slaughter.

Halal (Muslim) slaughter performed with a knife that is too short causes definite distress and struggling in cattle. We recommend to those Muslim religious authorities that require slaughter without stunning that the knife be razor sharp with a straight blade that is at least twice the width of the neck. Unless the animals are stunned, the use of curved skinning knives is not acceptable. Due to the fact that Muslim slaughtermen do not usually receive as extensive special training in slaughter techniques as Jewish shochtim, preslaughter stunning is strongly recommended. As stated earlier, reversible head-only electrical stunning is accepted by some Muslim religious authorities. Preslaughter stunning allows plants to run at higher line speeds and maintain high standards of animal welfare.

In some ritual slaughter plants animal welfare is compromised when animals are pulled out of the restraint box before they have lost sensibility. Observations clearly indicated that disturbance of the incision or allowing the cut edges to touch caused the animal to react strongly. Dragging the cut incision of a sensible animal against the bottom of the head opening device is likely to cause pain. **Animals must remain in the restraint device with the head holder and body restraint loosened until they collapse.** The belly lift should remain up during bleedout to prevent bumping of the incision against the head opening when the animal collapses.

Since animals cannot communicate, it is impossible to completely rule out the possibility that a correctly made incision may cause some unpleasant sensation. However, one can definitely conclude that poor cutting methods and stressful restraint methods are not acceptable. Poor cutting technique often causes vigorous struggling. When the cut is done correctly, behavioral reactions to the cut are much less than reactions to air hissing, metal clanging noises, inversion or excessive pressure applied to the body. Discomfort during a properly done shechitah cut is probably minimal because cattle will stand still and do not resist a comfortable head restraint device. Observations in many plants indicate that slaughter without stunning requires greater management attention to the details of the procedures than stunning in order to maintain good welfare. Ritual slaughter is a procedure that can be greatly improved by the use of a total quality management (TQM) approach to continual incremental improvements in the process. In plants with existing upright restraint equipment significant improvements in animal welfare and reductions in petechial haemorrhages can be made by making the following changes:

- Training of employees in gentle calm cattle handling
- Modifying the restrainer per the specifications in this article
- Eliminating distractions which make animals balk
- Careful attention to the exact cutting method

There needs to be continual monitoring and improvements in technique to achieve rapid onset of insensibility. A high incidence of prolonged sensibility is caused by poor cutting technique, rough handling, excessive pressure applied by the restraint device, or agitated excited animals (Grandin and Regenstein, 1994).

Note: The full text of the paper, the figures that accompany this paper, and the references cited therein can be found on Dr. Grandin's web site.

In summary, Dr. Grandin addresses the following key points with respect to the relationship between these kosher/halal guidelines and the AMI guidelines. Dr. Grandin has pointed out that for kosher/halal slaughter some of the key issues that are properly covered in the AMI guidelines are the importance of not abusing prods — the animal needs to be calm when entering the slaughter equipment. Also, the issue of slipping and falling because of improper flooring and the "spooking" of animals because of noise, improper lighting, items on the runway, etc., also need special attention as the special needs of kosher/halal slaughter need an even quieter/calmer animal than for regular slaughter. Dr. Grandin has also indicated that the current "vocalization score" does need to be raised slightly for kosher/halal from 3 to 5%, with the expectation that some cows vocalize while the head holder is being used.

LISTING OF THE FINAL GUIDELINES FOR RELIGIOUS SLAUGHTER

The information established through the two articles has been consolidated into a single set of actual guidelines for religious slaughter.

Equipment Requirements

Kosher and halal slaughter will be performed while the animal is in a comfortable upright position. Handling systems that turn the animal upside down or hang the animal are considered unacceptable. (This section was prepared by Dr. Temple Grandin.)

The ASPCA Pen and Other Box-Type Restrainers That Hold a Single Bovine or Other Animal That Weighs Over 400 lb (180 kg)

Plants using an ASPCA pen or other box-type restrainer must operate the equipment in such a manner that the animals are calm during the procedure, and slaughter takes place immediately after the head holder is in place. In pens equipped with a belly lift or other lifting devices that do not fully support the entire body, the animal must not be lifted off the floor. Lifting devices that support the entire body in a comfortable upright position are acceptable.

The ASPCA pen and all other box-type restrainers shall be operated at less than 100 animals per hour. This applies to animals weighing over 400 lb (180 kg). The rear pusher gate on the ASPCA pen and all other box-type restrainers must be equipped with a pressure limiting device.

Conveyor Restrainer Systems for All Species

The guideline applies to all systems where animals are restrained on one or more moving conveyors. The most common systems are the V-type restrainers, where the animals are held between two conveyors that form a V, or the center track (double rail), where the animals straddle a moving conveyor. Systems that use conveyors must be designed so that the animal's body is fully supported in a comfortable upright position.

To reduce balking at the conveyor entrance, installation of a false floor to prevent the incoming animals from seeing the "visual cliff" under an elevated conveyor is strongly recommended.

To prevent incoming animals from seeing people and other distractions through the discharge end of the conveyor, a solid barrier may need to be installed. Plant layout will determine the need for such a barrier.

Conveyor slats or flites must be designed to avoid pinching of the animal between the slats or flites. Most animals must ride quietly on the conveyor system with a minimum of struggling. Conveyor systems may be operated at speeds greater than 100 animals per hour.

Small Animal Systems

Small animals such as sheep, goats, and calves weighing under 400 lb (180 kg) can be handled in small versions of the ASPCA pen or other box-type restrainers or conveyor restrainers. All guidelines for these systems also apply to small animals. Small animals such as sheep and goats may be restrained in a standing upright position by one or more persons.

For small plants that ritually slaughter a few animals per week, a simple inexpensive restrainer device constructed from pipe can be used to hold the animal in a comfortable upright position. Devices that lift the animal off the floor are permitted only if they support the entire animal's body. One acceptable design is two parallel pipes that the animal straddles. The restrainer must hold most animals with a minimum of struggling.

Requirements for All Restraining Systems

The device must hold the animal firmly enough to provide the "feeling of restraint," but excessive pressure that causes struggling or vocalization must be avoided. Pressure limiting devices are required.

For cattle, the equipment should be operated in such a manner that the AMI guidelines of no more than 5 cattle out of 100 vocalizing is adhered to. Equipment must be designed to prevent jerky motion of all parts of the device that contact the animal. Steady, nonjerky motion of restraint devices helps to keep animals calm.

All restraint devices must use the concept of optimal (*not* maximum) pressure. The device must hold the animal firmly enough to provide the feeling of restraint but not so tight as to cause pain or discomfort. Struggling is a sign of discomfort.

All equipment must be engineered to reduce noise such as air hissing and metal-to-metal clanging and rattling. Some methods of noise reduction that should be used are muffling devices for air hissing and nonmetallic guides for moving metal parts. Locating hydraulic pumps and other noisy power units away from the restrainer is strongly recommended.

A solid barrier should be installed around the animal's head to prevent the animal from seeing people and other distractions in its flight zone. All equipment must be free of sharp edges and pinch points. Surfaces such as the neck openings must be rounded and smooth.

Moving parts of restraint equipment such as rear pusher gates, head holders, or other body restraint devices must be designed so that the operator has control and can incrementally apply pressure. The controls must provide the operator with the capability of easily stopping the moving parts at mid point positions to accommodate different sized animals.

Restraint equipment must be illuminated to encourage animals to enter and reduce balking. The entrance into all types of restraint equipment must be designed so that balking is reduced and the equipment will be able to comply with AMI guidelines on electric prod use.

Flooring both in the restraint equipment and at its entrance must comply with AMI guidelines on slipping and falling. Distractions that cause balking must be eliminated.

Head Holders

Head holders should be designed taking into account the materials found in Grandin and Regenstein (1994). To prevent excessive bending of the neck, the bovine's forehead should be parallel to the floor. All headholders must be equipped with pressure limiting devices.

Slaughter must take place immediately after the head holder is in place. The actual time between restraint of the head and actual slaughter must be less than 10 sec.

The head holder must hold the incision open during and immediately after the cut, but excessive stretching of the neck that could cause tearing of the incision must be avoided.

Knife Requirements

The FMI/NCCR Animal Welfare Committee recommends that a knife similar to the kosher chalef be used for halal slaughter. The knife should have a straight blade and be at least twice the width of the neck. It shall be kept properly sharpened at all times.

Slaughter

Careful attention to the exact cutting method used is necessary for each animal.

Postslaughter Requirements

Animals must remain in the restraint device with the head holder and body restraint loosened until they collapse. Animals should become insensible in less than a minute.

Employees

All employees must be trained in gentle, calm cattle handling including all stages of the slaughter

Based on this information, we propose that the following information be gathered as the starting place for evaluating religious slaughter. Some are actual audit points, whereas others are done to obtain information so that the audit forms can be refined in the future.

The actual audit points for religious slaughter can be found on the Animal Welfare Audit Program documents for "Non-Poultry" slaughter and for "Poultry Slaughter" accessed through www.fmi.org or www.nccr.net.

REFERENCES

Grandin, T. 1991. *Recommended Animal Handling Guide for Meat Packers,* 2nd ed., American Meat Institute, Washington, D.C.

Grandin, T. 1994. Euthanasia and slaughter of livestock. *J. Am. Vet. Med. Assoc.,* 204, 1354-1360.

Grandin, T. 1996. Factors that impede movement at slaughter plants, *J. Am. Vet. Med. Assoc.,* 209, 757-759.

Grandin, T. and Regenstein, J.M. 1994. Religious slaughter and animal welfare: a discussion for meat scientists, *Meat Focus Int.,* 3, 115-123.

Appendix L

KEY TERMINOLOGY FROM OTHER LANGUAGES

Note: [1]Arabic; [2]Hebrew; [3]Urdu.

Ahadith[1]	Plural of hadith
Ahlul Kitab[1]	People of the book, the ones who have received the scripture.
Allah[1]	The one God.
Allahu Akbar[1]	The God is greater.
As-Salat[1]	Daily and other prayers by a Muslim.
Bismillah[1]	In the name of God.
Bodek[2]	Inspector of internal organs.
Chalef[2]	A special knife for kosher kill.
Chometz[2]	Prohibited grains in Jewish laws.
Dhabh[1]	Islamic method of slaughtering an animal by slicing the neck.
Dhabiha[1]	Slaughtered according to the Islamic method.
Dhakaat[1]	Purification of meat by proper slaughtering of an animal.
Hadith[1]	Saying of Muhammad.
Haj[1]	Pilgrimage to Mecca by a Muslim.
Halaal/Halal[1]	Lawful or permitted in Islamic laws.
Halacha[2]	System of Jewish laws.
Haraam/Haram[1]	Unlawful or prohibited in Islamic laws.
Istihala[1]	Distinct chemical change in a product.
Jalalah/Jalallah[1]	Condition of an animal, eating filth and living in filth.
Janaba[1]	A condition of being unclean.
Khamr[1]	Fermented, frothing, alcoholic drinks.
Kosher[2]	Lawful or permitted in Jewish laws.

Makrooh[1]	Disliked or detested.
Mashbooh[1]	Doubtful or questionable.
Nahr[1]	Islamic method of slaughtering an animal by spearing or stabbing on the neck.
Najis[1]	Unclean or dirty.
Pareve[2]	A neutral nondairy, nonmeat kosher product.
Quran[1]	The Muslim holy scripture.
Shari'ah[1]	The Islamic jurisprudence or laws.
Shirk[1]	Ascribing partners to God.
Shochet[2]	A Jewish slaughter man.
Sunnah[1]	Traditions (sayings and actions) of Muhammad.
Tasmiyyah[1]	Invoking the name of God.
Torah[2]	The Jewish holy scripture.
Treife[2]	Not kosher or prohibited in the Jewish law.
Umra[1]	Short version of haj.
Zabh[3]	A variant pronunciation of dhabh.
Zabiha/Zabeeha[3]	A variant pronunciation of dhabiha.

Appendix M

E-NUMBERED INGREDIENTS

Source: Modified from Table 6 of Bender, D.A. and Bender, A.E., *Bender's Dictionary of Nutrition and Food Technology,* 7th ed., 1999, Woodhead Publishing, Cambridge, and CRC Press, Boca Raton.

Permitted Food Additives in the European Union[a]

E-Number	Additive	Halal Status

Coloring Materials: Used to Make Food More Colorful and Attractive or to Replace Color Lost in Processing

Yellow and Orange Colors

E-Number	Additive	Halal Status
E-100	Cucurmin	Halal
E-101	Riboflavin, riboflavin phosphate (Vitamin B$_2$)	Halal
E-102	Tartrazine (= FD&C Yellow No. 6)	Halal
E-104	Quinoline yellow	Halal
107	Yellow 2G	Halal
E-110	Sunset yellow FCF or orange yellow S (= FD&C Yellow No. 6)	Halal

Red Colors

E-Number	Additive	Halal Status
E-120	Cochineal or carminic acid	Doubtful
E-122	Carmoisine or azorubine	Halal
E-123	Amaranth	Halal
E-124	Ponceau 4R or cochineal red A	Doubtful
E-127	Erythrosine BS (= FD&C Red No. 3)	Halal
E-128	Red 2G	Halal
E-129	Allura red (= FD&C Red No. 40)	Halal

Blue Colors

E-Number	Additive	Halal Status
E-131	Patent blue V	Halal
E-132	Indigo carmine or indigotine (= FD&C Blue No. 2)	Halal
E-133	Brilliant blue FCF (= FD&C Blue No. 1)	Halal

Green Colors

E-Number	Additive	Halal Status
E-140	(i) Chlorophylls, the natural green color of leaves, (ii) chlorphyllins	Halal
E-141	Copper complexes of (i) chlorophylls, (ii) chlorophyllins	Halal
E-142	Green S or acid brilliant green BS	Halal

Brown and Black Colors

E-Number	Additive	Halal Status
E-150a	Plain caramel (made from sugar in the kitchen)	Halal
E-150b	Caustic sulfite caramel	Halal
E-150c	Ammonia caramel	Halal

Permitted Food Additives in the European Union[a] (continued)

E-Number	Additive	Halal Status
Brown and Black Colors (continued)		
E-150d	Sulfite ammonia caramel	Halal
E-151	Black PN or brilliant black BN	Halal
E-153	Carbon black or vegetable carbon (charcoal)	Halal
E-154	Brown FK	Halal
E-155	Brown HT (Chocolate brown HT)	Halal
Derivatives of Carotene		
E-160(a)	(i) Mixed carotenes, (ii) β-carotene	Halal
E-160(b)	Annatto, bixin, norbixin	Halal
E-160(c)	Paprika extract, capsanthin or capsorubin	Halal
E-160(d)	Lycopene	Halal
E-160(e)	β-Apo-8′-carotenal (Vitamin A active)	Halal
E-160(f)	Ethyl ether of β-apo-8′-carotenoic acid	Halal
Other Plant Colors		
E-161(a)	Flavoxanthin	Halal
E-161(b)	Lutein	Halal
E-161(c)	Cryptoxanthin	Halal
E-161(d)	Rubixanthin	Halal
E-161(e)	Violaxanthin	Halal
E-161(f)	Rhodoxanthin	Halal
E-161(g)	Canthaxanthin	Halal
E-162	Beetroot red or betanin	Halal
E-163	Anthocyanins (the pigments of many plants)	Halal
Inorganic Compounds Used as Colors		
E-170	(i) Calcium carbonate (chalk), (ii) calcium hydrogen carbonate	Halal
E-171	Titanium dioxide	Halal
E-172	Iron oxides and hydroxides	Halal
E-173	Aluminium	Halal
E-174	Silver	Halal
E-175	Gold	Halal
E-180	Pigment rubine or lithol rubine BK	Halal

Permitted Food Additives in the European Union[a] (continued)

E-Number	Additive	Halal Status

Preservatives: Protect Foods Against Microbes that Cause Spoilage and Food Poisoning; Compounds Increase the Safe Storage Life of Foods

Sorbic Acid and Its Salts

E-200	Sorbic acid	Halal
E-201	Sodium sorbate	Halal
E-202	Potassium sorbate	Halal
E-203	Calcium sorbate	Halal

Benzoic Acid and Its Salts

E-210	Benzoic acid (occurs naturally in many fruits)	Halal
E-211	Sodium benzoate	Halal
E-212	Potassium benzoate	Halal
E-213	Calcium benzoate	Halal
E-214	Ethyl p-hydroxybenzoate	Halal
E-215	Ethyl p-hydroxybenzoate, sodium salt	Halal
E-216	Propyl p-hydroxybenzoate	Halal
E-217	Propyl p-hydroxybenzoate, sodium salt	Halal
E-218	Methyl p-hydroxybenzoate	Halal
E-219	Methyl p-hydroxybenzoate, sodium salt	Halal

Sulfur Dioxide and Its Salts

E-220	Sulfur dioxide (also used to prevent browning of raw peeled potatoes)	Halal
E-221	Sodium sulfite	Halal
E-222	Sodium hydrogen sulfite	Halal
E-223	Sodium metabisulfite	Halal
E-224	Potassium metabisulfite	Halal
E-226	Calcium sulfite	Halal
E-227	Calcium hydrogen sulfite	Halal
E-228	Potassium hydrogen sulfite	Halal

Biphenyl and Its Derivatives

E-230	Biphenyl or diphenyl (for surface treatment of citrus fruits)	Halal
E-231	Orthophenylphenol (2-Hydroxybiphenyl) (for surface treatment of citrus fruits)	Halal
E-232	Sodium orthophenylphenol (sodium biphenyl-2-yl oxide)	Halal

Permitted Food Additives in the European Union[a] (continued)

E-Number	Additive	Halal Status
Other Preservatives		
E-233	2-(Thiazol-4-yl) benzimidazole (thiobendazole) (for surface treatment of citrus fruits and bananas)	Halal
E-234	Nisin	Halal
E-235	Natamycin (NATA) (for surface treatment of cheeses and dried cured sausages)	Halal
E-239	Hexamethylene tetramine (hexamine)	Halal
E-242	Dimethyl dicarbonate	Halal
E-912	Montan acid esters (for surface treatment of citrus fruits)	Doubtful
E-914	Oxidized polyethylene wax (for surface treatment of citrus fruits)	Halal
Pickling Salts		
E-249	Potassium nitrite	Halal[b]
E-250	Sodium nitrite	Halal[b]
E-251	Sodium nitrate	Halal[b]
E-252	Potassium nitrate (saltpetre)	Halal[b]
Acids and Their Salts: Used as Flavorings and as Buffers to Control the Acidity of Foods, in Addition to Their Antimicrobial Properties		
E-260	Acetic acid	Halal
E-261	Potassium acetate	Halal
E-262	(i) Sodium acetate, (ii) sodium hydrogen acetate (sodium diacetate)	Halal
E-263	Calcium acetate	Halal
E-270	Lactic acid	Halal
E-280	Propionic acid	Halal
E-281	Sodium propionate	Halal
E-282	Calcium propionate	Halal
E-283	Potassium propionate	Halal
E-284	Boric acid (as preservative in caviar)	Halal
E-285	Sodium tetraborate (borax) (as preservative in caviar)	Halal
E-290	Carbon dioxide	Halal
E-296	Malic acid	Halal
E-297	Fumaric acid	Halal

Permitted Food Additives in the European Union[a] (continued)

E-Number	Additive	Halal Status

Antioxidants: Used to Prevent Fatty Foods Going Rancid, and to Protect Fat-Soluble Vitamins (A, D, E, And K) Against the Damaging Effects of Oxidation

Vitamin C and Derivatives

E-Number	Additive	Halal Status
E-300	L-Ascorbic acid (Vitamin C)	Halal
E-301	Sodium-L-ascorbate	Halal
E-302	Calium- L-ascorbate	Halal
E-304	(i) Ascorbyl palmitate, (ii) ascorbyl stearate	Doubtful
E-315	Erythorbic acid (*iso*-ascorbic acid)	Halal
E-316	Sodium erythorbate (sodium *iso*-ascorbate)	Halal

Vitamin E

E-Number	Additive	Halal Status
E-306	Natural extracts rich in tocopherols	Halal
E-307	Synthetic α-tocopherol	Halal
E-308	Synthetic γ-tocopherol	Halal
E-309	Synthetic δ-tocopherol	Halal

Other Antioxidants

E-Number	Additive	Halal Status
E-310	Propyl gallate	Halal
E-311	Octyl gallate	Halal
E-312	Dodecyl gallate	Doubtful
E-320	Butylated hydroxyanisole (BHA)	Halal
E-321	Butylated hydroxytoluene (BHT)	Halal
E-322	Lecithins	Doubtful

More Acids and Their Salts: Used as Flavorings and as Buffers to Control the Acidity of Foods, in Addition to Other Special Uses

Salts of Lactic Acid (E-270)

E-Number	Additive	Halal Status
E-325	Sodium lactate	Halal
E-326	Potassium lactate	Halal
E-327	Calcium lactate	Halal
E-585	Ferrous lactate	Halal

Citric Acid and Its Salts and Esters

E-Number	Additive	Halal Status
E-330	Citric acid (formed in the body, and present in many fruits)	Halal
E-331	(i) Monosodium citrate, (ii) disodium citrate, (iii) trisodium citrate	Halal

Permitted Food Additives in the European Union[a] (continued)

E-Number	Additive	Halal Status
Citric Acid and Its Salts and Esters (continued)		
E-332	(i) Monopotassium citrate, (ii) dipostassium citrate, (iii) tripotassium citrate	Halal
E-333	(i) Monocalcium citrate, (ii) dicalcium citrate, (iii) tricalcium citrate	Halal
E-1505	Triethyl citrate	Halal
Tartaric Acid and Its Salts		
E-334	L(+)Tartaric acid (tartaric acid occurs naturally; as well as their properties as acids, tartrates are often used as sequestrants and emulsifying agents)	Halal
E-335	(i) Monosodium tartrate, (ii) disodium tartrate	Halal
E-336	(i) Monopotassium tartrate (cream of tartar), (ii) dipotassium tartrate	Halal
E-337	Sodium potassium tartrate	Halal
Phosphoric Acid and Its Salts		
E-338	Phosphoric acid	Halal
E-339	(i) Monosodium phosphate, (ii) disodium phosphate, (iii) trisodium phosphate	Halal
E-340	(i) Monopotassium phosphate, (ii) dipotassium phosphate, (iii) tripotassium phosphate	Halal
E-341	(i) Monocalcium phosphate, (ii) dicalcium phosphate, (iii) tricalcium phosphate	Halal
E-450	Diphosphates: (i) disodium diphosphate, (ii) trisodium diphosphate, (iii) tetrasodium diphosphate, (iv) dipotassium diphosphate, (v) tetrapotassium diphosphate, (vi) dicalcium diphosphate, (vii) calcium dihydrogen diphosphate	Halal
E-451	Triphosphates: (i) pentasodium triphosphate, (ii) pentapotassium triphosphate	Halal

Permitted Food Additives in the European Union[a] (continued)

E-Number	Additive	Halal Status
Phosphoric Acid and Its Salts (continued)		
E-452	Polyphosphates: (i) sodium polyphosphate, (ii) potassium polyphosphate, (iii) sodium calcium polyphosphate, (iv) calcium polyphosphate	Halal
E-540	Dicalcium diphosphate	Halal
E-541	Sodium aluminum phosphate, acidic	Halal
E-542	Edible bone phosphates (bone meal, used as anticaking agent)	Doubtful
E-544	Calcium polyphosphates (used as anticaking agent)	Halal
E-545	Ammonium polyphosphates (used as anticaking agent)	Halal
Salts of Malic Acid (E-296)		
E-350	Sodium malate	Halal
E-351	Potassium malate	Halal
E-352	Calcium malate	Halal
Other Acids and Their Salts		
E-353	Metatartaric acid	Halal
E-354	Calcium tartrate	Halal
E-355	Adipic acid	Halal
E-356	Sodium adipate	Halal
E-357	Potassium adipate	Halal
E-363	Succinic acid	Halal
E-370	1,4-Heptonolactone	Halal
E-375	Nicotinic acid	Halal
E-380	Triammonium citrate	Halal
E-381	Ammonium ferric citrate	Halal
E-385	Calcium disodium EDTA	Halal

Emulsifiers and Stabilizers: Used to Enable Oils and Fats to Mix with Water, to Give a Smooth and Creamy Texture to Food, and Slow the Staling of Baked Goods; Also Used to Make Jellies

Alginates		
E-400	Alginic acid (derived from seaweed)	Halal
E-401	Sodium alginate	Halal
E-402	Postassium alginate	Halal

Permitted Food Additives in the European Union[a] (continued)

E-Number	Additive	Halal Status
Alginates (continued)		
E-403	Ammonium alginate	Halal
E-404	Calcium alginate	Halal
E-405	Propane-1,2-diol alginate	Halal
Other Plant Gums		
E-406	Agar (derived from seaweed)	Halal
E-407	Carrageenan (derived from the seaweed Irish moss)	Halal
E-410	Locust bean gum (carob gum)	Halal
E-412	Guar gum	Halal
E-413	Tragacanth	Halal
E-414	Gum acacia (gum arabic)	Halal
E-415	Xanthan gum	Halal
E-416	Karaya gum	Halal
E-417	Tara gums	Halal
E-418	Gellan gums	Halal
Fatty Acid Derivatives		
E-430	Polyoxyethylene (8) stearate	Doubtful
E-431	Polyoxyethylene (40) stearate	Doubtful
E-432	Polyoxyethylene (20) sorbitan monolaurate (Polysorbate 20)	Doubtful
E-433	Polyoxyethylene (20) sorbitan mono-oleate (Polysorbate 80)	Doubtful
E-434	Polyoxyethylene (20) sorbitan monopalmitate (Polysorbate 40)	Doubtful
E-435	Polyoxyethylene (20) sorbitan monostearate (Polysorbate 60)	Doubtful
E-436	Polyoxyethylene (20) sorbitan tristearate (Polysorbate 65)	Doubtful
Pectin and Derivatives		
E-440	(i) Pectin, (ii) amidated pectin (pectin occurs in many fruits, and is often added to jam to help it set)	Halal
Other Compounds		
E-322	Lecithins	Doubtful
E-442	Ammonium phosphatides	Doubtful
E-444	Sucrose acetate isobutyrate	Doubtful
E-445	Glycerol esters of wood rosins	Doubtful

Permitted Food Additives in the European Union[a] (continued)

E-Number	Additive	Halal Status
Cellulose and Derivatives		
E-460	(i) Microcrystalline cellulose, (ii) powdered cellulose	Halal
E-461	Methyl cellulose	Halal
E-463	Hydroxypropyl cellulose	Halal
E-464	Hydroxypropylmethyl cellulose	Halal
E-465	Ethylmethyl cellulose	Halal
E-466	Carboxymethylcellulose, sodium carboxymethylcelluslose	Halal
Salts or Esters of Fatty Acids		
E-470a	Sodium, potassium, and calcium salts of fatty acids	Doubtful
E-470b	Magnesium salts of fatty acids	Doubtful
E-471	Mono- and diglycerides of fatty acids	Doubtful
E-472a	Acetic acid esters of mono- and diglycerides of fatty acids	Doubtful
E-472b	Lactic acid esters of mono- and diglycerides of fatty acids	Doubtful
E-472c	Citric acid esters of mono- and diglycerides of fatty acids	Doubtful
E-472d	Tartaric acid esters of mono- and diglycerides of fatty acids	Doubtful
E-472e	Mono- and diacetyl tartaric esters of mono- and diglycerides of fatty acids	Doubtful
E-472f	Mixed acetic and tartaric acid esters of mono- and diglycerides of fatty acids	Doubtful
E-473	Sucrose esters of fatty acids	Doubtful
E-474	Sucroglycerides	Doubtful
E-475	Polyglycerol esters of fatty acids	Doubtful
E-476	Polyglycerol esters of polycondensed esters of castor oil (polyglycerol polyricinoleate)	Doubtful
E-477	Propane-1,2-diol esters of fatty acids	Doubtful
E-478	Lactylated fatty acid esters of glycerol and propane-1,2-diol	Doubtful
E-479b	Thermally oxidized soybean oil interacted with mono- and diglycerides of fatty acids	Doubtful
E-481	Sodium stearoyl-2-lactylate	Doubtful

Permitted Food Additives in the European Union[a] (continued)

E-Number	Additive	Halal Status
Salts or Esters of Fatty Acids (continued)		
E-482	Calcium stearoyl-2-lactylate	Doubtful
E-483	Stearyl tartrate	Doubtful
E-491	Sorbitan monostearate	Doubtful
E-492	Sorbitan tristearate	Doubtful
E-493	Sorbitan monolaurate	Doubtful
E-494	Sorbitan monooleate	Doubtful
E-495	Sorbitan monopalmitate	Doubtful
E-1518	Glyceryl triacetate (triacetin)	Doubtful

Acids and Salts Used for Special Purposes: Buffers, Emulsifying Salts, Sequestrants, Stabilizers, Raising Agents, and Anticaking Agents

E-Number	Additive	Halal Status
Carbonates		
E-500	(i) Sodium carbonate, (ii) sodium bicarbonate (sodium hydrogen carbonate), (iii) sodium sesquicarbonate	Halal
E-501	(i) Potassium carbonate, (ii) potassium bicarbonate (potassium hydrogen carbonate)	Halal
E-503	(i) Ammonium carbonate, (ii) ammonium hydrogen carbonate	Halal
E-504	(i) Magnesium carbonate, (ii) magnesium hydrogen carbonate (magnesium hydroxide carbonate)	Halal
Hydrochloric Acid and Its Salts		
E-507	Hydrochloric acid (ordinary salt is sodium chloride)	Halal
E-508	Potassium chloride (sometimes used as a replacement for ordinary salt)	Halal
E-509	Calcium chloride	Halal
E-510	Ammonium chloride	Halal
E-511	Magnesium chloride	Halal
E-512	Stannous chloride	Halal
Sulfuric Acid and Its Salts		
E-513	Sulfuric acid	Halal
E-514	(i) Sodium sulfate, (ii) sodium hydrogen sulfate	Halal

Permitted Food Additives in the European Union[a] (continued)

E-Number	Additive	Halal Status
Sulfuric Acid and Its Salts (continued)		
E-515	(i) Potassium sulfate, (ii) potassium hydrogen sulfate	Halal
E-516	Calcium sulfate	Halal
E-517	Ammonium sulfate	Halal
E-518	Magnesium sulfate	Halal
E-520	Aluminium sulfate	Halal
E-521	Aluminium sodium sulfate	Halal
E-522	Aluminium potassium sulfate	Halal
E-523	Aluminium ammonium sulfate	Halal
Alkalis: Used as Bases to Neutralize Acids in Foods		
E-524	Sodium hydroxide	Halal
E-525	Potassium hydroxide	Halal
E-526	Calcium hydroxide	Halal
E-527	Ammonium hydroxide	Halal
E-528	Magnesium hydroxide	Halal
E-529	Calcium oxide	Halal
E-530	Magnesium oxide	Halal
Other Salts		
E-535	Sodium ferrocyanide	Halal
E-536	Potassium ferrocyanide	Halal
E-538	Calcium ferrocyanide	Halal
E-540	Dicalcium diphosphate	Halal
E-541	Sodium aluminium phosphate, acidic	Halal
Compounds Used as Anticaking Agents, and for Other Uses		
E-542	Edible bone phosphate (bone meal)	Doubtful
E-544	Calcium polyphosphates	Halal
E-545	Ammonium polyphosphates	Halal
Silicon Salts		
E-551	Silicon dioxide (silica, sand)	Halal
E-552	Calcium silicate	Halal
E-553a	(i) Magnesium silicate, (ii) magnesium trisilicate	Halal
E-553b	Talc	Halal
E-554	Sodium aluminium silicate	Halal
E-555	Potassium aluminium silicate	Halal
E-556	Calcium aluminium silicate	Halal

Permitted Food Additives in the European Union[a] (continued)

E-Number	Additive	Halal Status
Other Compounds		
E-558	Bentonite	Halal
E-559	Kaolin (aluminium silicate)	Halal
E-570	Fatty acids	Doubtful
E-572	Magnesium stearate	Doubtful
E-574	Gluconic acid	Halal
E-575	Glucono-δ-lactone	Halal
E-576	Sodium gluconate	Halal
E-577	Potassium gluconate	Halal
E-578	Calcium gluconate	Halal
E-579	Ferrous gluconate	Halal
E-585	Ferrous lactate	Halal
Compounds Used as Flavor Enhancers		
E-620	ʟ-Glutamic acid (a natural amino acid)	Doubtful
E-621	Monosodium glutamate (MSG)	Doubtful
E-622	Monopotassium glutamate	Doubtful
E-623	Calcium diglutamate	Doubtful
E-624	Monoammonium glutamate	Doubtful
E-625	Magnesium diglutamate	Doubtful
E-626	Guanylic acid	Halal
E-627	Disodium guanylate	Halal
E-628	Dipotassium guanylate	Halal
E-629	Calcium guanylate	Halal
E-630	Inosinic acid	Halal
E-631	Disodium inosinate	Halal
E-632	Dipotassium inosinate	Halal
E-633	Calcium inosinate	Halal
E-634	Calcium 5'-ribonucleotides	Halal
E-635	Disodium 5-ribonucleotides	Halal
E-636	Maltol	Halal
E-637	Ethyl maltol	Halal
E-640	Glycine and its sodium salt (a natural amino acid)	Doubtful
E-900	Dimethylpolysiloxane	Halal
Compounds Used as Glazing Agents		
E-901	Beeswax	Halal
E-902	Candelilla wax	Halal
E-903	Carnauba wax	Halal

Permitted Food Additives in the European Union[a] (continued)

E-Number	Additive	Halal Status
Compounds Used as Glazing Agents (continued)		
E-904	Shellac	Doubtful
E-912	Montan acid esters	Doubtful
E-914	Oxidized polyethylene wax	Halal
Compounds Used to Treat Flour		
E-920	L-Cysteine hydrochloride (a natural amino acid)	Doubtful
924	Potassium bromate	Halal
925	Chlorine	Halal
926	Clorine dioxide	Halal
927	Azodicarbamide	Halal
Propellant Gases		
E-938	Argon	Halal
E-939	Helium	Halal
E-941	Nitrogen	Halal
E-942	Nitrous oxide	Halal
E-948	Oxygen	Halal
Sweeteners and Sugar Alcohols		
E-420	(i) Sorbitol, (ii) sorbitol syrup	Halal
E-421	Mannitol	Halal
E-422	Glycerol	Doubtful
E-927a	Azodicarbonamide	Halal
E-927b	Carbamide	Halal
E-950	Acesulfame K	Halal
E-951	Aspartame	Doubtful
E-952	Cyclamic acid and its sodium and calcium salts	Halal
E-953	Isomalt	Halal
E-954	Saccharine and its sodium, potassium, and calcium salts	Halal
E-957	Thaumatin	Halal
E-959	Neohesperidin didihydrochalcone	Halal
E-965	(i) Maltitol, (ii) maltitol syrup	Halal
E-966	Lactitol	Halal
E-967	Xylitol	Halal

Permitted Food Additives in the European Union[a] **(continued)**

E-Number	Additive	Halal Status
Miscellaneous Compounds		
E-999	Quillaia extract	Halal
E-1105	Lysozyme	Doubtful
E-1200	Polydextrose	Halal
E-1201	Polyvinyl pyrrolidone	Halal
E-1202	Polyvinyl polypyrrolidone	Halal
E-1505	Triethyl citrate	Halal
E-1518	Glyceryl triacetate (triacetin)	Doubtful
Modified Starches		
E-1404	Oxidized starch	Halal
E-1410	Monostarch phosphate	Halal
E-1412	Distarch phosphate	Halal
E-1413	Phosphated distarch phosphate	Halal
E-1414	Acetylated distarch phosphate	Halal
E-1420	Acetylated starch	Halal
E-1422	Acetylated starch adipate	Halal
E-1440	Hydroxypropyl starch	Halal
E-1442	Hydroxypropyl distarch phosphate	Halal
E-1450	Starch sodium octanoyl succinate	Halal

Note: Those without the prefix E are permitted in U.K. but not throughout the E.U.

[a] The status of ingredients given here is for the pure form. Many of the ingredients are standardized with other food ingredients, status of which may be doubtful, making the listed ingredient doubtful. Ask the supplier for a full disclosure of the composition.

[b] Due to their effect on health, some Islamic scholars consider these preservatives as doubtful.

Appendix N

HALAL STATUS
FOR INGREDIENTS

Halal Status for Ingredients

Ingredient Name	Description	Halal Status
Acesulfame potassium	A synthetic sweetener	Halal
Acetone peroxide	A dough conditioner and maturing and bleaching agent	Halal
Acetylated monoglyceride	An emulsifier manufactured by the interesterification of edible fats with triacetin	Doubtful
Acidophilus	A bacterial starter culture; used to develop flavor	Halal
Adipic acid	An acidulant and flavoring agent	Halal
Agar or agar-agar	A gum obtained from red seaweeds	Halal
Albumin	Any of several water-soluble proteins from egg white, blood serum, and milk	Doubtful
Algin and alginates	A gum; a term used for derivatives of alginic acid, which are obtained from brown seaweed	Halal
All-purpose flour	A flour that is intermediate between long-patent flours (bread flour and cake flour)	Halal
Allspice	A spice made from the dried, nearly ripe berries of *Pimenta officinalis*, a tropical evergreen tree	Halal
Almond oil	The oil of the bitter almond after the removal of hydrocyanic acid	Halal
Almond paste	A paste made by cooking sweet and bitter almonds which have been ground and blanched in combination with sugar	Halal
Alum	A preservative; the inclusive term for several aluminum-type compounds such as aluminum sulfate and aluminum potassium sulfate	Halal
Aluminum acetate	A general-purpose food additive	Halal
Aluminum ammonium sulfate	A general-purpose food additive that functions as a buffer and neutralizing agent	Halal
Aluminum calcium silicate	An anticaking agent; used in vanilla powder	Halal
Aluminum caprate	The aluminum salt of capric acid	Doubtful
Aluminum caprylate	The aluminum salt of caprylic acid	Doubtful

Halal Status for Ingredients (continued)

Ingredient Name	Description	Halal Status
Aluminum laurate	The aluminum salt of lauric acid	Doubtful
Aluminum myristate	The aluminum salt of myristic acid	Doubtful
Aluminum nicotinate	The aluminum salt of nicotinic acid	Halal
Aluminum oleate	The aluminum salt of oleoic acid	Doubtful
Aluminum oxide	A dispersing agent	Halal
Aluminum palmitate	The aluminum salt of palmitic acid	Doubtful
Aluminum sodium sulfate	A general-purpose food additive that functions as a buffer, neutralizing agent, and firming agent	Halal
Aluminum stearate	The aluminum salt of stearic acid	Doubtful
Aluminum sulfate	A starch modifier and firming agent	Halal
Amidated pectin	The low-methoxyl pectin resulting from deesterification with ammonia	Halal
Ammonium alginate	A gum that is the ammonium salt of alginic acid	Halal
Ammonium bicarbonate	A double strengthener and leavening agent	Halal
Ammonium carbonate	A leavening agent and pH control agent that consists of ammonium bicarbonate and ammonium carbamate	Halal
Ammonium caseinate	The ammonium salt of casein	Halal
Ammonium chloride	A dough conditioner and yeast food	Halal
Ammonium hydroxide	An alkaline, clear, colorless solution of ammonia	Halal
Ammonium phosphate	A general-purpose food additive	Halal
Ammonium sulfate	A dough conditioner, firming agent, and processing aid	Halal
Ammonium sulfite	An additive; used in the production of caramel	Halal
Anise	A spice that is the dried, ripe fruit of *Pimpinella anisum,* a small herb	Halal
Annatto	A color source of yellowish to reddish-orange color obtained from the seed coating of the tree *Bixa orellanna*	Halal

Halal Status for Ingredients (continued)

Ingredient Name	Description	Halal Status
Arabic gum	A gum obtained from breaks or wounds in the bark of *Acacia* trees	Halal
Arabinogalactan	A gum, the plant extract obtained from larch trees	Halal
Arginine	A nonessential amino acid that exists as white crystals or powder	Doubtful
Arrowroot	A starch obtained from *Maranta arundinacea*, a perennial that produces starchy rhizomes	Halal
Artificial coloring	Any of the FD&C synthetic coloring materials	Doubtful
Artificial flavor	Any substance whose function it is to impart flavor and that is not derived from a spice, fruit, vegetable, edible yeast, bark, bud, herb, root, leaf, or other plant material, meat, seafood, eggs, poultry, dairy, or fermentation products	Doubtful
Ascorbic acid	It is termed Vitamin C, a water-soluble vitamin that prevents scurvy, helps maintain the body's resistance to infection, and is essential for healthy bones and teeth	Halal
Ascorbyl palmitate	An antioxidant formed by combining ascorbic acid with palmitic acid	Doubtful
Ascorbyl stearate	An antioxidant; used in peanut oil	Doubtful
Aspartame	A synthetic sweetener that is a dipeptide, synthesized by combining the methyl ester of phenylalanine with aspartic acid	Doubtful
Aspartic acid	A nonessential amino acid that exists as colorless or white crystals of acid taste	Doubtful
Azodicarbonamide	A dough conditioner that exists as a yellow to orange-red crystalline powder	Halal
Babassu oil	The oil obtained from the nut of the babassu palm	Halal

Halal Status for Ingredients (continued)

Ingredient Name	Description	Halal Status
Baker's yeast	Dried microorganism *Saccharomyces cerevisiae*	Halal
Baker's yeast glycan	The dried cell walls of yeast, *Saccharomyces cerevisiae*; used as an emulsifier and thickener in salad dressing	Halal
Baking powder	A leavening agent that consists of a mixture of sodium bicarbonate, one or more leavening agents such as sodium aluminum phosphate or monocalcium phosphate, and an inert material such as starch	Halal
Baking soda	Bicarbonate of soda, chemically known as sodium bicarbonate	Halal
Barley	A cereal grain of which there are winter and spring types; used in malting of barley	Halal
Basil	A spice obtained from the dried leaves and tender stems of *Ocimum basilicum L.*	Halal
Bay leaves	A spice flavoring that consists of the dried leaves obtained from the evergreen tree *Laurus nobilis*, also called sweet bay or laurel	Halal
Beeswax	The purified wax obtained from the honeycomb of the bee; used to glaze candy, and in chewing gum and confections	Halal
Beet extract	A natural red colorant obtained from beets; used in yogurt, beverages, candies, and desserts	Halal
Bentonite	A general-purpose additive; used as a pigment and colorant	Halal
Benzoic acid	A preservative that occurs naturally in some foods such as cranberries, prunes, and cinnamon	Halal
Benzoyl peroxide	A colorless, crystalline solid with a faint odor of benzaldehyde resulting from the interaction of benzoyl chloride and a cooled sodium peroxide solution; used in specified cheeses	Halal

Halal Status for Ingredients (continued)

Ingredient Name	Description	Halal Status
Beta-apo-8'-catotenal	A colorant that is a carotenoid producing a light to dark orange hue	Halal
Beta-carotene	A colorant that is a carotenoid producing a yellow to orange hue	Halal
Biotin	A water-soluble vitamin that is a nutrient and dietary supplement	Doubtful
Birch	An artificial flavoring; used in soft drinks	Halal
Bixin	A carotenoid that is the main coloring component of annatto; obtained from the *Bixa orellana* tree	Halal
Bleached flour	Flour that has been whitened by the removal of the yellow pigment; bleaching can be obtained during natural aging of the flour or can be accelerated by chemicals that are usually oxidizing agents	Halal
Bran	The seed husk or outer coatings of cereals, such as wheat, rye, and oats that is separated from the flour	Halal
Bread flour	A hard-wheat flour, which generally is obtained from straight or long patent flours	Halal
Bromated flour	A white flour to which potassium bromate is added; used in baked goods	Halal
Bromelin	A proteolytic enzyme obtained from pineapple; used in tenderizing beef	Halal
Brominated vegetable oil (BVO)	A vegetable oil whose density has been increased to that of water by being combined with bromine; flavoring oils are dissolved in the brominated oil, which can then be added to fruit drinks	Halal

Halal Status for Ingredients (continued)

Ingredient Name	Description	Halal Status
Brown sugar	A sweetener that consists of sucrose crystals covered with a film of cane molasses which gives it the characteristic color and flavor	Halal
Buckwheat	A member of the grass family that is usually marketed as a cereal grain; buckwheat grain	Halal
Bulgur	A precooked cracked wheat that retains the bran and germ fraction of the grain	Halal
Butter	A source of milk fat obtained from cream by a churning process which results in the production of butter fat and buttermilk	Halal
Butter, clarified	Butter that has undergone purification by the removal of solid particles or impurities that may affect the color, odor, or taste	Halal
Buttermilk	The product that remains when fat is removed from milk or cream in the process of churning into butter; cultured buttermilk is prepared by souring buttermilk	Halal
Buttermilk, dried	The powder form of buttermilk	Halal
Butter oil	The clarified fat portion of milk, cream, or butter obtained by the removal of the nonfat constituent	Halal
Butylated hydroxyanisole (BHA)	An antioxidant that imparts stability to fats and oils	Halal
Butylated hydroxytoluene (BHT)	An antioxidant that functions similarly to butylated hydroxyanisole (BHA) but is less stable at high temperatures	Halal
Butyric acid	A fatty acid that is commonly obtained from butter fat	Halal
Caffeine	A white powder or needles that are odorless and have a bitter taste, occurring naturally in tea leaves, coffee, cocoa, and cola nuts	Halal

Halal Status for Ingredients (continued)

Ingredient Name	Description	Halal Status
Cake flour	A soft wheat flour	Halal
Calciferol	A fat-soluble vitamin; termed vitamin D2	Halal
Calcium acetate	The calcium salt of acetic acid which functions as a sequestrant and mold-control agent	Halal
Calcium alginate	The calcium salt of alginic acid which functions as a stabilizer and thickener	Halal
Calcium aluminum silicate	An anticaking agent that permits the free flow of dry ingredients	Halal
Calcium ascorbate	The salt of ascorbic acid which is a white to slightly yellow crystalline powder	Halal
Calcium bromate	A dough conditioner, maturing, and bleaching agent which exists as a white crystalline powder	Halal
Calcium caprate	The calcium salt of capric acid; used as a binder, emulsifier, anticaking agent, and as a general additive	Doubtful
Calcium caprylate	The calcium salt of caprylic acid; used as a binder, emulsifier, anticaking agent, and as a general additive	Doubtful
Calcium carbonate	The calcium salt of carbonic acid; used as an anticaking agent and dough strengthener	Halal
Calcium caseinate	The calcium salt of casein	Halal
Calcium chloride	A general-purpose food additive	Halal
Calcium citrate	The calcium salt of citric acid; used as a sequestrant, buffer, and firming agent	Halal
Calcium diacetate	The salt of acetic acid; used as a preservative and sequestrant	Halal
Calcium gluconate	A white crystalline powder; used as a firming agent, formulation aid, sequestrant, and stabilizer	Halal
Calcium glycerophosphate	A nutrient and dietary supplement; used in baking powder	Halal

Halal Status for Ingredients (continued)

Ingredient Name	Description	Halal Status
Calcium hydroxide	A general food additive made by adding water to calcium oxide (lime)	Halal
Calcium hydroxyphosphate	A food additive; used in table salt	Halal
Calcium iodate	Calcium salt of iodine; used as a dough conditioner in bread	Halal
Calcium lactate	The calcium salt of lactic acid	Halal
Calcium lactobionate	The calcium salt of lactobionic acid; used as a firming agent in dry pudding mixes	Halal
Calcium laurate	The calcium salt of lauric acid; used as a binder, emulsifier, and anticaking agent	Doubtful
Calcium metaphosphate	A sequestrant; used as a general additive	Halal
Calcium myristate	The calcium salt of myristic acid; used as a binder, emulsifier, and anticaking agent	Doubtful
Calcium oleate	The calcium salt of oleic acid; used as a binder, emulsifier, and anticaking agent	Doubtful
Calcium oxide	A general food additive	Halal
Calcium palmitate	The calcium salt of palmitic acid; used as a binder, emulsifier, and anticaking agent	Doubtful
Calcium pantothenate	A nutrient and dietary supplement; used in special dietary foods	Halal
Calcium pectinate	The salt of pectin which is obtained from citrus or apple fruit	Halal
Calcium peroxide	A dough conditioner; used in bakery products	Halal
Calcium phosphate	A phosphate existing in several forms in nature; used as an anticaking agent and mineral supplement	Halal
Calcium phytate	A chelating agent; used to bind metallic ions in foods to prevent discoloration and off-flavor	Halal

Halal Status for Ingredients (continued)

Ingredient Name	Description	Halal Status
Calcium propionate	The salt of propionic acid; used as a preservative	Halal
Calcium pyrophosphate	A nutrient and dietary supplement	Halal
Calcium saccharate	A derivative of calcium hydroxide and sugar; used to improve the whipping ability of whipping cream	Halal
Calcium saccharin	A sweetener that is the calcium form of saccharin; ca. 500 times sweeter than sucrose	Halal
Calcium silicate	An anticaking agent; used in baking powder	Halal
Calcium sorbate	The calcium salt of sorbic acid; used in cheese	Halal
Calcium stearate	The calcium salt of stearic acid; used as an anticaking agent, binder, and emulsifier	Doubtful
Calcium stearoyl lactylate	The calcium salt of lactic acid and stearic acid; used as a dough conditioner, whipping agent, and emulsifier	Doubtful
Calcium sulfate	A general additive made by the high-temperature calcining of gypsum	Halal
Canthaxanthin	A synthetic red colorant similar to carotenoids	Halal
Caramel color	A dark-brown color obtained by controlled oxidation of starch and other carbohydrates	Halal
Carbonated water	A beverage made by absorbing carbon dioxide in water	Halal
Carbon dioxide	A gas obtained during fermentation of glucose; used in the carbonation of beverages	Halal
Carboxymethylcellulose (CMC)	A water-soluble gum cellulose; used as a thickener, stabilizer, binder, film former, and suspending agent	Halal
Cardamon	A spice from the dried, ripe seed of *Elettaria cardamomum*	Halal

Halal Status for Ingredients (continued)

Ingredient Name	Description	Halal Status
Carmine	The red colorant aluminum lake of carminic acid which is the coloring pigment obtained from dried bodies of the female insect *Coccus cacti*	Doubtful
Carnauba wax	A general-purpose additive obtained from leaf buds and leaves of the Brazilian wax palm *Copernicia cerifera*	Halal
Carob	A cocoa substitute obtained from the pods of the carob tree *Ceratonia siliqua*	Halal
Carotene	A colorant and provitamin; used in ice cream, cheese, and other dairy products	Halal
Carrageenan	A seaweed extract obtained from red seaweed	Halal
Casein	A milk protein which is prepared commercially from skim milk by the precipitation with lactic, hydrochloric, or sulfuric acid	Halal
Caseinates	Sodium or calcium salts of casein that are produced by neutralizing acid casein with calcium or sodium hydroxide	Halal
Celery seed	A spice made from the dried, ripe fruit of the herb *Apium graveolens*	Halal
Cellulose	A carbohydrate polymer; used as a fiber source and bulking agent in low-calorie foods	Doubtful
Cheese	The product obtained by the coagulation of the milk protein by suitable enzymes or bacteria	Doubtful
Cheese culture	Several bacteria; used in the coagulation of the milk protein	Halal
Chervil	A spice derived from the plant *Anthriscus cerefolium*	Halal
Chewing gum base	A formulation containing masticatory substances such as chicle and several other GRAS substances; used in the manufacture of chewing gum	Doubtful

Halal Status for Ingredients (continued)

Ingredient Name	Description	Halal Status
Chia oil	The oil from the seed of plants of the genus *Salvia*; used in the preparation of soft drinks	Halal
Chicle	A natural masticatory substance of vegetable origin; used in chewing gum base	Halal
Chilte	A substance of vegetable origin; used in chewing gum base	Halal
Chives	A spice from the *Allium* plant similar to green onions	Halal
Chloropentafluoro-ethane	A propellant and aerating agent for foamed or sprayed foods	Halal
Chlorophyll	A green pigment from plants; used in sausage casings and shortening	Halal
Cholic acid	An emulsifier; used as an emulsifying agent in egg white	Halal
Choline	A substance of the Vitamin B complex family	Doubtful
Cider vinegar	The product made by the alcoholic and subsequent acetic fermentation of apple juice; used in salad dressings, mayonnaise, and sauces	Halal
Cinnamon	A spice made from the dried bark of the evergreen tree *Cinnamomum cassia*	Halal
Citric acid	An acidulant and antioxidant produced by extraction from lemon and lime juice	Halal
Citrus oil	A flavorant obtained by pressing the oil from the rind of citrus fruits	Halal
Clove	A spice that is the unripened bud from the clove tree, *Eugenia caryophyllata*	Halal
Cochineal	A red colorant extracted from the dried bodies of the female insect *Coccus cacti*; the coloring is carminic acid in which the water-soluble extract is cochineal	Doubtful
Cocoa butter	The fat obtained by pressing chocolate liquor	Halal

Halal Status for Ingredients (continued)

Ingredient Name	Description	Halal Status
Cocoa liquor	Viscous liquid obtained by the grinding of cocoa nibs from the cocoa bean; also termed chocolate liquor	Halal
Cocoa powder	The powder produced by grinding cocoa presscake obtained from the cocoa nibs	Halal
Coconut	The fruit of coconut palm which produces coconut meat and coconut oil	Halal
Collagen	A protein that is the principal constituent of connective tissue; used in casings and personal-care products	Doubtful
Copper gluconate	Salt of gluconic acid; used as a dietary supplement	Halal
Copra	A dried coconut meat; used as a flavorant	Halal
Coriander	A spice that is the dried, ripe fruit of *Coriandrum sativum* L.	Halal
Corn bran	A dry-milled product of high fiber content obtained from corn	Halal
Corn flour	A finely ground flour made from milling and sifting maize or corn	Halal
Cornmeal	A ground corn of granular form	Halal
Corn oil	The oil obtained from the germ of the corn plant	Halal
Cornstarch	The starch made from the endosperm of corn	Halal
Cornstarch, modified	A starch produced by treating cornstarch with dilute mineral acid or other chemicals	Halal
Corn syrup	A corn sweetener liquid composed of maltose, dextrin, dextrose, and other polysaccharides	Halal
Corn syrup solids	The dry form of corn syrup	Halal
Cottonseed oil	The oil obtained from the seeds of the cotton plant	Halal
Cranberry extract	A natural red colorant from cranberries	Halal
Cream	Portion of milk that is high in milk fat, with 18 to 40% fat	Halal

Halal Status for Ingredients (continued)

Ingredient Name	Description	Halal Status
Cream of tartar	The potassium salt of tartaric acid; used in baked goods, candies, and puddings	Halal
Cumin	A spice that is the dried, ripe fruit of *Cuminum cyminum* L.	Halal
Cupric/cuprous aspartate	A salt of aspartic acid with copper	Doubtful
Cupric/cuprous carbonate	A copper salt of carbonic acid	Halal
Cupric/cuprous chloride	A copper salt	Halal
Cupric/cuprous citrate	A copper salt of citric acid	Halal
Cupric/cuprous glycerophosphate	A copper salt of glycerine and phosphate	Halal
Cupric/cuprous nitrate	A copper salt of nitric acid	Halal
Cupric/cuprous sulfate	A copper salt of sulfuric acid	Halal
Cuprous iodide	A salt of copper and iodine; used in table salt	Halal
Curry powder	A blend of spices; used as seasoning in curries, sauces, and meats; typical spices include coriander, ginger, clove, cinnamon, red pepper, and cumin	Halal
Cyanocobalamin	Vitamin B12, a water-soluble vitamin found in meat, fish, and milk	Doubtful
Cysteine	A nonessential amino acid; used to increase elasticity; can be made from human hair, duck feathers, or synthetically	Doubtful
Cystine	A nonessential amino acid; used as a nutrient and dietary supplement	Doubtful
Deoxycholic acid	An emulsifier; used in dried egg whites	Halal
Dextran	A gum obtained by fermentation of sugar with bacteria	Halal
Dextrin	A compound formed from acid or enzymes	Halal
Dextrose	A corn sweetener commercially made from starch	Halal

Halal Status for Ingredients (continued)

Ingredient Name	Description	Halal Status
Diacetyl tartaric acid esters of mono- and diglycerides	A hydrophilic emulsifier; used in oil-in-water emulsions	Doubtful
Diammonium phosphate	A leavening agent; used in the production of cookies and crackers	Halal
Dicalcium phosphate	A mineral supplement and dough conditioner	Halal
Diglyceride	An emulsifier prepared by esterification of two fatty acids with glycerol or by interesterification between glycerol and triglycerides	Doubtful
Dill seed	A spice from the dried, ripe fruit of the plant *Anethum graveolens* L.	Halal
Dill weed	A spice from the leaf of the dill plant	Halal
Dimethylpolysiloxane	An antifoaming agent; used in fats and oils	Halal
Dioctyl sodium sulfosuccinate	An emulsifiying agent; used as a flavor potentiator in canned milk	Doubtful
Dipotassium phosphate	The potassium salt of phosphoric acid; used as a stabilizing salt, buffer, and sequestrant	Halal
Disodium calcium EDTA	A sequestrant and chelating agent	Halal
Disodium dihydrogen EDTA	A sequestrant and chelating agent	Halal
Disodium 5'-guanylate (DSG)	A flavor enhancer that belongs to the same family as monosodium glutamate	Halal
Disodium 5'-inosinate	A flavor enhancer which performs as does disodium guanylate	Halal
Disodium malate	An acidulant and a sodium salt of malic acid	Halal
Disodium phosphate	The sodium salt of phosphoric acid; used as a protein stabilizer and mineral supplement	Halal
Distilled monoglyceride	An emulsifier containing a minimum of 90% monoglyceride derived from edible fat and glycerine	Doubtful

Halal Status for Ingredients (continued)

Ingredient Name	Description	Halal Status
Dodecyl gallate	An antioxidant; used in cream cheese, instant mashed potatoes, margarine, fats, and oils	Halal
Dough conditioner	A blend of minerals; used in baked goods	Halal
Durum flour	The fine powder obtained from durum wheat	Halal
EDTA (ethylene-diaminetetraacetate)	A sequestrant and chelating agent	Halal
Egg albumen	The protein fraction of egg, also termed egg white	Halal
Egg yolk	The yellow portion of the egg; used as an emulsifier in mayonnaise and salad dressing	Halal
Enriched bleach flour	Flour that has been whitened by removal of the yellow pigments and fortified with vitamins and minerals	Halal
Erythorbic acid	A food preservative that is a strong reducing agent	Halal
Ester gum	A density adjuster prepared from glycerol of nonanimal sources and refined wood rosin of pine trees	Halal
Ethoxylated mono- and diglycerides	An emulsifier prepared by the glycerolysis of edible vegetable fats and reacting with ethylene oxide	Doubtful
Ethoxyquin	An antioxidant; used in the preservation of color in chili powder, ground chili, and paprika	Halal
Ethyl maltol	A synthetic flavor enhancer related to maltol	Halal
Ethyl vanillin	A flavoring agent that is a synthetic vanilla flavor	Halal
Eugenol	A flavoring obtained from clove oil and also found in carnation and cinnamon leaves	Halal
Farina	Wheat granules from which the bran and germ have been removed	Halal

Halal Status for Ingredients (continued)

Ingredient Name	Description	Halal Status
Fat	Water-insoluble material of plant or animal origin, consisting of triglycerides that are semisolid at room temperature	Doubtful
Fatty acids	A mixture of aliphatic acids of plant or animal origin; fatty acids are used as lubricants, binders, food processing defoamers, and emulsifiers	Doubtful
FD&C blue # 1	A colorant, also called brilliant blue; used in candies, baked goods, soft drinks, and desserts	Halal
FD&C blue # 2	A colorant, also called indigotine; used in candies, confections, and baked goods	Halal
FD&C green # 3	A colorant, also called fast green FCF; used in cereals, soft drinks, beverages, and desserts	Halal
FD&C red # 3	A colorant, also called erythrocine; used in candies, confections, and cherry dyeing	Halal
FD&C red # 4	A colorant, also called ponceau SX; used only in maraschino cherries	Halal
FD&C red # 40	A colorant, also called Allura red AC; used in beverages, desserts, candies, confections, cereals, and ice cream	Halal
FD&C yellow # 5	A colorant, also called tartrazine; used in beverages, baked goods, pet foods, desserts, candies, confections, cereal, and ice cream	Halal
FD&C yellow # 6	A colorant, also called sunset yellow FCF; used in beverages, bakery goods, dessert confections, and ice cream	Halal
Fennel	A spice that is the dried, ripe fruit of the plant *Foeniculum vulgare* Mil.	Halal
Fenugreek	Seeds of the herb *Trigonella foenumgraecum*; used as a spice and flavoring	Halal

Halal Status for Ingredients (continued)

Ingredient Name	Description	Halal Status
Ferric ammonium citrate	A nutrient and dietary supplement that is a source of iron, containing 17% iron	Halal
Ferric chloride	A nutrient and dietary supplement	Halal
Ferric/ferrous ammonium sulfate	A nutrient and dietary supplement	Halal
Ferric/ferrous sulfate	A nutrient and dietary supplement	Halal
Ferric fructose	A nutrient and dietary supplement	Halal
Ferric glycerophosphate	A nutrient and dietary supplement	Halal
Ferric nitrate	A nutrient and dietary supplement	Halal
Ferric orthophosphate	A nutrient and dietary supplement	Halal
Ferric oxide	A nutrient and dietary supplement	Halal
Ferric pyrophosphate	A nutrient and dietary supplement	Halal
Ferrous carbonate	A nutrient and dietary supplement	Halal
Ferrous citrate	A nutrient and dietary supplement	Halal
Ferrous fumarate	A nutrient and dietary supplement	Halal
Ferrous gluconate	A nutrient and dietary supplement and a coloring adjunct	Halal
Ferrous lactate	A nutrient and dietary supplement	Halal
Ferrous sulfate	A nutrient and dietary supplement	Halal
Ferrous tartrate	A nutrient and dietary supplement	Halal
Flavoring	Any ingredients, natural or artificial, single or in a mixture, that impart flavor to a food, components of a flavoring generally not revealed to the consumer	Doubtful
Folic acid	A B-complex vitamin found in liver, nuts, and green vegetables	Doubtful
Fructose (fruit sugar)	A sweetener found naturally in fresh fruit and honey; obtained by the inversion of sucrose with the enzyme invertase and by the isomerization of dextrose	Halal
Fructose corn syrup	A corn sweetener derived from the isomerization of glucose in the syrup to fructose by the enzyme isomerase	Halal
Fumaric acid	An acidulant; used in desserts, pie fillings, and candies	Halal

Halal Status for Ingredients (continued)

Ingredient Name	Description	Halal Status
Furcellaran	A gum extract of the red alga *Furcellaria fastigiata*; used in milk puddings, flans, jelly, jam, and other products	Halal
Garlic	A spice that is cloves of the herb *Allium sativium*	Halal
Garlic salt	A seasoning that is a mixture of garlic powder and salt	Halal
Gelatin	A protein of animal origin that functions as a gelling agent; obtained from collagen derived from beef bones and calf skin, pork skin, fish skin, or poultry skin	Doubtful
Ghatti	A gum that is a plant exudate from the *Anogeissus latifolia* tree; used in buttered syrup and as a stabilizer	Halal
Ginger	A spice that is the dried rhizome of the ginger plant, *Zingiber officinale*	Halal
Glacial acetic acid	A strong acidulant preservative flavoring	Halal
Gluconic acid	A mild organic acid which is the hydrolyzed form of glucono-delta-lactone	Halal
Glucono-delta-lactone (GDL)	A mild acidulant; used in sausages, frankfurters, and dessert mixes	Halal
Glutamic acid	An amino acid; used as a flavor enhancer, nutrient, dietary supplement, and salt substitute	Doubtful
Glutamic acid hydrochloride	An amino acid; used as a flavoring agent	Doubtful
Glycerin or glycerol	A polyol; used as a humectant, crystallization modifier, and plasticizer in candies, baked goods, and other products	Doubtful
Glycerol ester of wood rosin	Same as ester gum	Halal
Glyceryl-lacto esters of fatty acids	Emulsifiers that are the lactic acid esters of mono- and diglycerides; used as emulsifiers and plasticizers in toppings, cakes, and icings	Doubtful

Halal Status for Ingredients (continued)

Ingredient Name	Description	Halal Status
Glycerol-lacto-stearate	An emulsifier that is a monoglyceride esterified with lactic acid; used in whipped toppings, shortenings, cake mixes, and coatings	Doubtful
Glyceryl monolaurate	A monoglyceride emulsifier produced by the esterification of glycerine and lauric acid; used in baked goods, whipped toppings, frosting, and glazes	Doubtful
Glyceryl triacetate	A triglyceride of acetic acid; used as a humectant and solvent	Doubtful
Glycine	A nonessential amino acid; used as a nutrient and dietary supplement in artificially sweetened soft drinks	Doubtful
Glycoholic acid	An emulsifier; used in dried egg whites	Halal
Glycyrrhizin	A flavorant and foaming agent derived from licorice root	Halal
Graham flour	See whole wheat flour	Halal
Grain vinegar	An acidulant, also called distilled or spirit vinegar, made by the fermentation of dilute distilled alcohol	Halal
Grape color extract	A solution of grape pigment made from Concord grapes; used for coloring nonbeverage foods	Halal
Grape seed oil	The oil obtained from grape seeds	Halal
Grape skin extract	A neutral red colorant; used in soft drinks and candies	Halal
Guaiac gum	An antioxidant and preservative made from wood resin	Halal
Guar	A gum obtained from the seed kernel of the guar plant, *Cyamopsis tetragonoloba*; a thickener and stabilizer; used in ice cream, baked goods, and sauces	Halal

Halal Status for Ingredients (continued)

Ingredient Name	Description	Halal Status
Gum base	The component of chewing gum that is insoluble in water and remains after chewing; prepared from several ingredients such as chicle, crown gum, petroleum wax, lanolin, polythylene, polyvinyl acetate, rubber, paraffin, and antioxidants	Doubtful
Heptyl paraben	A preservative and antimicrobial agent	Halal
High-fructose corn syrup (HFCS)	A sweetener made from corn syrup through isomerization of the glucose in the syrup to fructose by the enzyme isomerase	Halal
Honey	A sweetener that is a natural syrup, made by honey bees, through the action of the enzyme honey invertase on nectar gathered by bees	Halal
Horseradish	A spice, obtained from the horseradish plant; used in sauces	Halal
Hydrochloric acid	An inorganic acid; used as an acidulant and neutralizing agent	Halal
Hydrogenated vegetable oil	Oil that has been hydrogenated to modify the texture from a liquid to a semisolid or solid through the chemical addition of hydrogen	Halal
Hydrolyzed vegetable protein (HVP)	A flavor enhancer obtained from vegetable proteins such as wheat gluten, corn gluten, defatted soy flour through acid hydrolysis; used to improve flavors in soups, dressings, meats, snack foods, and crackers	Halal
Hydroxylated soybean lecithin	An emulsifier and clouding agent obtained by treating soybean lecithin with peroxide; used in dry-mixed beverages, margarine, and baked goods	Halal

Halal Status for Ingredients (continued)

Ingredient Name	Description	Halal Status
Hydroxypropyl cellulose	A gum obtained from the reaction of alkali cellulose with propylene oxide; used in whipped toppings as a stablilizing and foaming aid	Halal
Hydroxypropyl methylcellulose	A gum formed by the reaction of propylene oxide and methyl chloride with alkali cellulose; used in bakery goods, dressings, breaded foods, and salad dressings	Halal
Invert sugar	A sweetener that is a mixture of dextrose (glucose) and fructose; used in candies and icings	Halal
Iron, reduced	A nutrient; used in cereals	Halal
Isoamyl butyrate	A synthetic flavoring agent; used in dessert gels, puddings, and baked goods	Halal
Isoamyl formate	A synthetic flavoring agent; used in dessert gels, puddings, candies, and ice cream	Halal
Isoamyl hexanoate	A synthetic flavoring agent; used in desserts, candies, and ice cream	Halal
Isobutyl cinnamate	A synthetic flavoring agent; used in beverages, ice creams, candies, and baked goods	Halal
Isobutyl formate	A synthetic flavoring agent; used in beverages, ice creams, candies, and baked goods	Halal
Isopropyl citrate	An antioxidant made by reacting citric acid with isopropyl alcohol; used in vegetable oils	Halal
Karaya	A gum, the dried exudate from the *Sterculia urens* tree; used in baked goods, denture adhesives, toppings, and frozen desserts	Halal
Kelp	A brown seaweed product; used as a source of iodine and as a flavor enhancer	Halal
Kola nut	The seed of kola species; used in beverages as a flavoring	Halal
L-Cysteine	See cysteine	Doubtful

Halal Status for Ingredients (continued)

Ingredient Name	Description	Halal Status
ʟ-Glutamine	An amino acid; isolated from sugarbeet juice; can be detected in most plants and animals (including bacteria)	Doubtful
ʟ-Taurine	An amino acid, usually sythentic	Doubtful
Lactic acid	An acidulant that is a natural organic acid present in milk; used as a flavor agent, preservative, and acidity adjuster in foods	Halal
Lactose	A milk sugar that occurs in most mammalian milk; usually obtained from cow milk; used in baked goods for flavor, browning, and tenderizing and in dry mixes as an anticaking agent	Halal
Lactylated fatty acid esters of glycerol and propylene glycol	An emulsifier made by the reaction of propylene glycol ester with lactic acid; used in whipped toppings and coffee whiteners	Doubtful
Lard	A fat rendered from hogs, consisting principally of oleic and palmitic fatty acids; used in many food and nonfood products	Haram
Lauric acid	A fatty acid usually from coconut oil and other vegetable fats; used as a lubricant, binder, and defoaming agent	Doubtful
Leavening agents	Mild acids that chemically react with alkaline sodium bicarbonate to produce carbon dioxide gas; leavening agents include tartaric acid, monocalcium phosphate, sodium acid pyrophosphate, sodium aluminum phosphate, and acidic acid	Halal
Lecithin	An emulsifier that is a mixture of phosphatides commercially obtained from soybeans and egg yolk; extensively used in foods	Halal
Lemon oil	An oil obtained from lemon fruit; used to impart lemon flavor	Halal

Halal Status for Ingredients (continued)

Ingredient Name	Description	Halal Status
Licorice extract	A flavoring agent made from dried root portions of *Glycyrrhiza glabra*; used in candies, baked goods, beverages, and tobacco	Halal
Limonene	An antioxidant and flavoring agent obtained from citrus oils	Halal
Locust bean gum	A gum obtained from the plant seed of a locust bean tree, *Ceratonia siligua*; used in processed cheese, ice cream, bakery products, soups, and pies	Halal
Mace	A spice comprising skin covering of the nutmeg *Myristica fragrans* Houtt.	Halal
Magnesium carbonate	An anticaking agent and general-purpose food additive	Halal
Magnesium caseinate	The magnesium salt of caseinate; used in bakery goods, drinks, and dietary supplements	Halal
Magnesium chloride	A dietary supplement and food additive	Halal
Magnesium hydroxide	A general-purpose food additive	Halal
Magnesium laurate	The magnesium salt of lauric acid; used as an emulsifier and anticaking agent	Doubtful
Magnesium myristate	The magnesium salt of myristic acid; used as emulsifier and anticaking agent	Doubtful
Magnesium oleate	The magnesium salt of oleic acid; used as an emulsifier and anticaking agent	Doubtful
Magnesium oxide	A nutrient and dietary supplement	Halal
Magnesium palmitate	The magnesium salt of palmitic acid; used as an emulsifier and anticaking agent	Doubtful
Magnesium silicate	An insoluble salt; used as an anticaking agent	Halal
Magnesium stearate	The magnesium salt of stearic acid; used as a lubricant, binder, emulsifier, and anticaking agent	Doubtful
Magnesium sulfate	A nutrient and dietary supplement	Halal

Halal Status for Ingredients (continued)

Ingredient Name	Description	Halal Status
Maleic acid	An organic acid; used as a preservative for fats and oils	Halal
Malic acid	An acid mainly from apples; used in soft drinks, dry-mix beverages, puddings, jellies, and fruit filling	Halal
Malt	A source of the enzyme alpha-amylase which hydrolyzes starch to fermentable sugars such as dextrins and maltose, produced by the controlled sprouting of grains, usually barley; used as a supplement to flour	Halal
Malt barley flour	The barley produced under the controlled sprouting of the barley grain and milling	Halal
Malt extract	An extract of water-soluble enzymes from barley evaporated to form a concentrate that contains D-alpha-amylase enzyme	Halal
Malt flour	The flour prepared by the drying and grinding of barley or wheat sprouted under controlled conditions	Halal
Maltodextrin	A product obtained from the partial acid or enzymatic hydrolysis of starch; used in crackers, puddings, and candies	Halal
Maltol	A flavor enhancer or modifier; occurs naturally in chicory, cocoa, coffee, and cereals; used in vanilla and chocolate-flavored foods and beverages	Halal
Maltose	A sugar formed by the enzymatic action of enzymes on starch; used in bread, instant foods, and pancake syrups	Halal
Malt syrup	The syrup obtained from barley by extraction and evaporation; used as a malt flavor, source of malt and protein, and is used in bakery goods	Halal

Halal Status for Ingredients (continued)

Ingredient Name	Description	Halal Status
Malt vinegar	A vinegar made by the alcoholic and subsequent acetous fermentation of malted barley or cereals; used as an acidulant and preservative in foods	Halal
Manganese carbonate	A source of manganese; used as a nutrient and dietary supplement	Halal
Manganese chloride	A nutrient and dietary supplement	Halal
Manganese oxide	A nutrient and dietary supplement	Halal
Manganese sulfate	A nutrient and dietary supplement	Halal
Mannitol	A polyhydric alcohol; used as a sweetener, humectant, and bulking agent in sugarless candies, chewing gum, cereal, and pressed mints	Halal
Maple sugar	A sweetener obtained by concentrating the sap of the maple sugar tree; used in syrups and candies	Halal
Maple syrup	A sweetener made by concentrating the sap of the sugar maple tree; used in syrups and candies	Halal
Margarine	A butter-like substance made by emulsifying oils and milk; vegetable oils or mixtures of vegetable oils and animal fat might be used	Doubtful
Marjoram	A spice obtained from leaves of the herb *Majorana bortensis* Moench.	Halal
Methylcellulose	A gum composed of cellulose; used in coatings and pie fillings	Halal
Methylethylcellulose	A gum that is the methyl ether of ethyl cellulose; used in whipped toppings, meringues, and aerated confectionary products	Halal
Methylparaben	An antimicrobial agent; used to prevent yeast and mold growth	Halal
Methyl salicylate (wintergreen oil)	A synthetic flavoring agent; used in chewing gum, candies, beverages, and baked goods	Halal

Halal Status for Ingredients (continued)

Ingredient Name	Description	Halal Status
Microcrystalline cellulose	A water insoluble gum; used in tablets, capsules, and shredded cheese as a non-nutritive filler, and anticaking agent	Halal
Milk fat or butter fat	The fat of milk concentrated as cream to make butter; used in bakery products, confections, and frozen desserts	Halal
Milk powder	The dry, whole milk that is produced by a spray or roller-drying process; used in soup mixes, dessert mixes, and for reconstituting	Halal
Milk solids-not-fat	The dry form of skim milk; used in ice cream mix, baked goods, and desserts	Halal
Mint	A spice derived from the mint family	Halal
Modified starch	Starch treated with certain chemicals to modify the physical characteristics of the native starch for improved solubility and texture; used as a thickener, binder, and stabilizer	Halal
Molasses	The by-product of the manufacture of sugar from sugar cane containing sucrose and invert sugar; used as a flavoring and sweetener	Halal
Monoammonium glutamate	A flavor enhancer obtained from glutamic acid; used in low-salt diets	Doubtful
Monoammonium phosphate	An acidulant; used as a leavening agent and yeast nutrient in bread	Halal
Mono- and diglycerides	A mixed emulsifier containing monoglycerides and diglycerides made by reacting glycerol with fats or oils; used in numerous food applications	Doubtful
Monocalcium phosphate	An acid-leavening agent and nutritional supplement; widely used in foods	Halal

Halal Status for Ingredients (continued)

Ingredient Name	Description	Halal Status
Monoglyceride	An emulsifier prepared by the direct esterification of fatty acids with glycerol or by the interesterification between glycerol and other triglycerides	Doubtful
Monoglyceride citrate	A sequestrant that is a mixture of glyceryl monooleate and citric acid; used as an antioxidant synergist	Doubtful
Monopotassium phosphate	A mild acid neutralizing agent and sequestrant; used in low-sodium products, milk, and meat products	Halal
Monosodium glutamate (MSG)	A flavor enhancer sodium salt of glutamic acid; used in meats, soups, and sauces as a flavor intensifier	Doubtful
Monosodium phosphate	An acidulant, buffer, and sequestrant; used in cheese and carbonated beverages	Halal
Mustard	A spice made from the dried, ripe seed of several varieties of the family Cruciferae; used as a flavorant in baked goods, sauces, and salad dressings	Halal
Mustard flour	The ground dehusked mustard seed; used in salad dressings and sauces and as a condiment	Halal
Myristic acid	A fatty acid obtained from coconut oil and other fats; used as a lubricant and defoaming agent	Doubtful
Natamycin	A preservative; used as a coating on the surface of cheeses to prevent the growth of mold or yeast	Halal
Niacin	A water-soluble B vitamin; used as a nutrient and dietary supplement	Halal
Niacinamide	A water-soluble B vitamin; used as a nutrient and dietary supplement	Halal

Halal Status for Ingredients (continued)

Ingredient Name	Description	Halal Status
Nitrous oxide	A noncombustible gas; used as a propellant in certain dairy and vegetable fat whipped toppings contained in pressurized containers; also called laughing gas	Halal
Nutmeg	A spice obtained from the nutmeg tree *Myristica fragrans*	Halal
Oat flour	Fine-mesh-ground dehulled oats	Halal
Oatmeal	Coarse-ground dehulled oats	Halal
Oleic acid	An unsaturated fatty acid; used as a lubricant and defoamer	Doubtful
Oleoresins	Spice extractions containing the volatile and nonvolatile flavor components; used in seasonings for foods	Halal
Oleoresin paprika	A colorant extraction containing the volatile and nonvolatile flavor components of paprika	Halal
Oxystearin	A modified fatty acid composed of glycerides of partially oxidized stearic and other fatty acids	Doubtful
Palmitic acid	A fatty acid composed principally of palmitic acid with varying amounts of stearic acid; used as a lubricant, binder, and defoaming agent	Doubtful
Palm kernel oil	An oil obtained from palm kernels; used interchangeably with coconut oil used in margarine and confectionary	Halal
Palm oil	The oil obtained from the fruit of the palm tree; used in margarine and shortenings	Halal
Pantothenic acid	A water-soluble B vitamin found in liver, eggs, and meat	Doubtful
Papain	A meat tenderizer that is a protein-digesting enzyme obtained from the papaya fruit	Halal
Paprika	A spice and colorant made from the ground, dried, ripe fruit of sweet red peppers	Halal

Halal Status for Ingredients (continued)

Ingredient Name	Description	Halal Status
Parabens	Antimicrobial agents that are methyl or propyl esters of para-hyroxybenzoic acid; used in baked goods, beverages, and food color	Halal
Parboiled rice	The rice that is cooked to be gelatinized, then dried; used in soups and rice dinners	Halal
Parsley	A herb made from the dried leaves of *Petroselinum hortense*; used for garnishing and seasoning	Halal
Pastry flour	A flour obtained from soft wheat	Halal
Peanut oil	The oil obtained from peanuts	Halal
Pectin	A water-soluble plant gum obtained from citrus peel and apple pomace	Halal
Pepper	A spice from the vine *Piper nigrum* L. which produces green, red, black and white berries	Halal
Pepper, cayenne	A hot spice related to paprika, bell peppers of the Capsicum family	Halal
Pepper, red	The ripe pod spice of the genus *Capsicum*	Halal
Petrolatum	A mixture of semisolid hydrocarbons obtained from petroleum; used in bakery products, dehydrated fruits and vegetables, and egg white solids as a release agent and defoamer	Halal
Petroleum wax	A wax from petroleum; used in chewing gum base and as a protective coating on raw fruits and vegetables	Halal
Phosphoric acid	A strong inorganic acid produced by burning phosphorus; used as a flavoring acid in cola and root beer beverages to provide desirable acidity	Halal
Polydextrose	A reduced-calorie bulking agent made by condensation of dextrose	Halal
Polyethylene glycol	Polymers of ethylene oxide and water; used as flavoring adjuvants in carbonated beverages	Halal

Halal Status for Ingredients (continued)

Ingredient Name	Description	Halal Status
Polyglycerol esters of fatty acids	Emulsifiers that are mixed partial esters formed by reacting polymerized glycerols with edible fats, oils, or fatty acids; used in cake mixes, whipped toppings, and in flavors and colors as a solubilizer	Doubtful
Polyoxyethylene sorbitan fatty acid esters	Emulsifiers made by reacting ethylene oxide with sorbitan esters; used in oil and water emulsions	Doubtful
Polyoxyethylene (20) sorbitan monooleate (Polysorbate 80)	An emulsifier produced by reacting oleic acid with sorbitol to yield a product which is reacted with ethylene oxide	Doubtful
Polyoxyethylene (20) sorbitan monostearate (Polysorbate 60)	An emulsifier manufactured by reacting stearic acid with sorbitol to yield a product which is reacted with ethylene oxide	Doubtful
Polyoxyethylene (20) sorbitan tristearate (Polysorbate 65)	An emulsifier manufactured by reacting stearic acid with sorbitol to yield a product which is then reacted with ethylene oxide	Doubtful
Polyoxyl (40) stearate	An emulsifier and antifoaming agent	Doubtful
Poppy seed	A seed spice of the *Papaver somniferum* L. plant	Halal
Potassium alginate	A gum that is the potassium salt of alginic acid	Halal
Potassium bicarbonate	An alkali and leavening agent	Halal
Potassium bisulfite	A preservative and an antioxidant	Halal
Potassium bromate	A dough conditioner	Halal
Potassium carbonate	A general-purpose food additive and alkali	Halal
Potassium chloride	A nutrient and dietary supplement used as a salt substitute	Halal
Potassium citrate, monohydrate	A sequestrant food additive	Halal
Potassium hydroxide	A water-soluble alkali and a food additive	Halal
Potassium iodate	A nutrient made by reacting iodine with potassium hydroxide	Halal
Potassium iodine	A nutrient and dietary supplement	Halal

Halal Status for Ingredients (continued)

Ingredient Name	Description	Halal Status
Potassium metabisulfite	A chemical preservative and antioxidant	Halal
Potassium nitrate	A preservative and color fixative in meats	Halal
Potassium nitrite	A color fixative in meats	Halal
Potassium oleate	The potassium salt of oleic acid; used as an emulsifier, and anticaking agent	Doubtful
Potassium palmitate	The potassium salt of palmitic acid; used as a binder, emulsifier and anticaking agent	Doubtful
Potassium sorbate	A preservative, the potassium salt of sorbic acid; used in cheese, bread, beverages, margarine, and dry sausage	Halal
Potassium stearate	The potassium salt of stearic acid; used as a platicizer in chewing gum base	Doubtful
Potassium sulfite	A preservative and antioxidant sulfite salt	Halal
Potassium tripolyphosphate	A phosphate; used as a moisture binder in meat and as an emulsifier and a sequestrant	Halal
Potato starch	A starch obtained from potatoes	Halal
Powdered sugar	A sweetener obtained by grinding sugar and adding cornstarch	Halal
Pregelatinized starch	Starch that has been heat processed to permit swelling in cold water; used in instant puddings	Halal
Propane	An aerating agent; used as a propellant and aerating agent for foamed or sprayed foods	Halal
Propionic acid	The acid source of the propionates; used as a mold inhibitor	Halal
Propylene glycol	A humectant and flavor solvent	Halal
Propylene glycol alginate	A gum that is the propylene glycol ester of alginic acid obtained from kelp	Halal
Propylene glycol mono- and diesters	An emulsifier that consists of propylene glycol esters of fatty acids such as palmitic and stearic	Doubtful

Halal Status for Ingredients (continued)

Ingredient Name	Description	Halal Status
Propylene glycol monostearate	An emulsifier that is a propylene glycol ester of stearic acid	Doubtful
Propyl gallate	A synthetic antioxidant; used to retard fat and oil rancidity	Halal
Pyridoxine	A water-soluble B vitamin found in liver, eggs, and meat	Doubtful
Pyridoxine hydrochloride	A water-soluble B vitamin	Doubtful
Quince seed	A gum produced from the fruit of the quince tree, *Cydonia oblonga*	Halal
Quinine	A flavorant naturally obtained from the cinchona tree; used as a bitter flavoring in beverages such as quinine water	Halal
Raisin	A dried grape; used as a fruit and as an ingredient in cereals	Halal
Rape seed oil	The oil derived from seeds of *Brassica campestris* and related plants	Halal
Rennet	A milk coagulant that is the concentrated extract of rennin enzyme obtained from calves' stomachs (calf rennet) or adult bovine stomachs (bovine rennet)	Doubtful
Riboflavin	The water-soluble vitamin B2	Halal
Rice bran oil	An oil extracted from rice bran	Halal
Rice bran wax	A refined wax obtained from rice bran	Halal
Rice flour	The flour made from different varieties of long-, medium-, and short-grain rice	Halal
Rice starch	The starch obtained from rice; used in puddings	Halal
Roselle	A natural red colorant obtained from the roselle flower extract from hibiscus	Halal
Rosemary	A herb made from the dried leaves of *Rosmarinus officinalis* L., an evergreen shrub	Halal
Rum ether (ethyl oxyhydrate)	A synthetic flavoring agent; used in beverages, candies, and ice cream	Halal
Rye flour	The flour obtained by milling rye	Halal

Halal Status for Ingredients (continued)

Ingredient Name	Description	Halal Status
Saccharin	A nonnutritive synthetic sweetener which is 300 to 400 times sweeter than sucrose	Halal
Safflower oil	A vegetable oil obtained from safflower seeds	Halal
Saffron	A spice and colorant obtained from the dried stigmas of *Crocus sativus* L.	Halal
Sage	A herb made from the dried leaves of the shrub *Salvia officinalis* L.	Halal
Sago starch	The starch obtained from the sago palm; used in puddings	Halal
Savory	A herb that is the dried leaves and flowering tops of the plant *Satureia hortensis* L.	Halal
Self-rising flour	White flour with added sodium bicarbonate and other salts	Halal
Semen cydonia	A product from quince seeds	Halal
Semolina	The purified coarse ground durum wheat	Halal
Sesame oil	The oil obtained from sesame seeds	Halal
Sesame seed	The seed of the plant *Sesamum indicum* L.	Halal
Shallot	A spice from *Allium ascalonicum*, a member of the onion family	Halal
Shortening	Any animal or vegetable fat or oil; used in baked goods	Doubtful
Silicon dioxide	An anticaking agent of mineral source	Halal
Smoke flavoring	A flavorant obtained from burning hardwoods; used and mixed with other ingredients	Doubtful
Sodium acetate	A source of acetic acid	Halal
Sodium acid pyrophosphate (SAPP)	A leavening agent, preservative, sequestrant, and buffer	Halal
Sodium alginate	A gum obtained as a sodium salt of alginic acid obtained from seaweed	Halal
Sodium aluminum phosphate	An emulsifier and leavening agent	Halal
Sodium aluminum sulfate	A leavening agent	Halal

Halal Status for Ingredients (continued)

Ingredient Name	Description	Halal Status
Sodium ascorbate	An antioxidant and a vitamin C	Halal
Sodium benzoate	A preservative that is the sodium salt of benzoic acid	Halal
Sodium bicarbonate	A leavening agent	Halal
Sodium bisulfate	A strong acid and food additive	Halal
Sodium bisulfite	A preservative that prevents discoloration and inhibits bacterial growth	Halal
Sodium calcium alginate	A gum that is the sodium and calcium salt of alginic acid; used as a thickener	Halal
Sodium calcium aluminosilicate	An anticaking agent	Halal
Sodium caprate	The sodium salt of capric acid; used as an emulsifier and anticaking agent	Doubtful
Sodium caprylate	The sodium salt of caprylic acid; used as an emulsifier and anticaking agent	Doubtful
Sodium carbonate	An alkali and food additive	Halal
Sodium caseinate	The sodium salt of casein, a milk protein	Halal
Sodium citrate	A buffer and sequestrant made from citric acid	Halal
Sodium diacetate	A preservative, sequestrant, acidulant, and flavoring agent	Halal
Sodium erythorbate	An antioxidant that is the sodium salt of erythorbic acid	Halal
Sodium hexametaphosphate	A sequestrant and moisture binder	Halal
Sodium hydroxide	An alkali and a neutralizing agent	Halal
Sodium iron pyrophosphate	A nutrient and dietary supplement	Halal
Sodium lactate	A humectant; the sodium salt of lactic acid	Halal
Sodium laurate	The sodium salt of lauric acid; used as a emulsifier	Doubtful
Sodium lauryl sulfate	An emulsifier and whipping aid	Doubtful
Sodium metabisulfite	A preservative and antioxidant	Halal
Sodium myristate	The sodium salt of myristic acid; used as a binder, emulsifier, and anticaking agent	Doubtful

Halal Status for Ingredients (continued)

Ingredient Name	Description	Halal Status
Sodium nitrate	The salt of nitric acid; used as an antimicrobial agent and preservative	Halal
Sodium nitrite	The salt of nitrous acid; used as an antimicrobial agent and preservative	Halal
Sodium oleate	The sodium salt of oleic acid; used as an emulsifier and anticaking agent	Doubtful
Sodium palmitate	The sodium salt of palmitic acid; used as an emulsifier and anticaking agent	Doubtful
Sodium polyphosphate	A sequestrant and emulsifier	Halal
Sodium potassium tartrate	A buffer and sequestrant	Halal
Sodium propionate	A preservative that is the salt of propionic acid	Halal
Sodium silicate	A preservative for eggs	Halal
Sodium silicoaluminate	An anticaking and conditioning agent used to prevent caking	Halal
Sodium sorbate	A preservative that is the salt of sorbic acid	Halal
Sodium stearate	The sodium salt of stearic acid; used as a plasticizer in chewing gum base	Doubtful
Sodium stearoyl fumarate	A dough conditioner for yeast-raised baked goods	Doubtful
Sodium stearyl fumarate	A dough conditioner for yeast-raised baked goods	Doubtful
Sodium tartrate	A sequestrant and stabilizer	Halal
Sodium tetrametaphosphate	A sequestrant and emulsifier	Halal
Sodium thiosulfate	A sequestrant and antioxidant	Halal
Sodium tripolyphosphate	A binder stabilizer and sequestrant	Halal
Sorbic acid	A preservative against yeasts and molds	Halal
Sorbitan monostearate	An emulsifier that is a sorbitan fatty acid ester, being a sorbitol-derived analog of glycerol monostearate	Doubtful
Sorbitol	A humectant and sugarless sweetener	Halal

Halal Status for Ingredients (continued)

Ingredient Name	Description	Halal Status
Soy flour	Flour obtained from defatted soybean	Halal
Soybean oil	The oil obtained from the seed of the soybean legume	Halal
Soybean protein concentrate	The concentrate obtained by processing soybean flour to remove soluble carbohydrates	Halal
Soybean protein isolate	The isolate prepared from soybean flour by extracting the protein	Halal
Spice	A variety of dried plant products that impart an aroma and flavor to foods	Halal
Stannous chloride	An antioxidant and preservative	Halal
Stearic acid	A fatty acid composed of a mixture of solid organic acids, principally stearic acid and palmitic acid	Doubtful
Stearoyl lactylate	A dough conditioner, emulsifier, and whipping agent	Doubtful
Stearyl citrate	An antioxidant made by reacting citric acid with stearyl alcohol	Doubtful
Succinic acid	An acidulant prepared by the hydrogenation of maleic or fumaric acid	Halal
Succinylated monoglycerides	Emulsifiers and dough conditioners	Doubtful
Sulfur dioxide	A preservative	Halal
Sulfuric acid	An acidulant	Halal
Sunflower oil	A vegetable oil obtained from sunflower seeds	Halal
Tallow	Animal fat from mutton or beef	Doubtful
Tarragon	The dried leaves and flowering tops of the herb *Artemisia dracunculus* L.	Halal
Tartaric acid	An acidulant and flavoring	Halal
Tertiary butylhydroquinone (TBHQ)	A synthetic antioxidant	Halal
Tetrasodium pyrophosphate	A coagulant, emulsifier, and a sequestrant	Halal
Textured soy flour	Soy flour that is heat processed and extruded	Halal

Halal Status for Ingredients (continued)

Ingredient Name	Description	Halal Status
Textured vegetable protein	A vegetable protein that is heat processed and extruded to form meat analogs	Halal
Thiamine	Water-soluble vitamin B1	Halal
Thyme	The herb composed of leaves of the plant *Thymus vulgaris* L.	Halal
Titanium dioxide	A white pigment; used as a color additive	Halal
Tocopherol	A fat-soluble vitamin E obtained from vegetable oils	Halal
Tofu	A soybean curd product	Halal
Tragacanth	A gum produced from the plant of the genus *Astragalus*; used in salad dressings and sauces	Halal
Tricalcium phosphate	An anticaking agent	Halal
Tricalcium silicate	An anticaking agent	Halal
Triethyl citrate	A sequestrant; used in lemon drinks	Halal
Trihydroxybutyro-phenone (THBP)	An antioxidant	Halal
Tripotassium citrate	A buffer and a sequestrant	Halal
Tripotassium phosphate	An emulsifier and alkaline buffer; used in low-sodium products	Halal
Trisodium citrate	A buffer and sequestrant	Halal
Trisodium phosphate	An emulsifier and buffer	Halal
Turbinado sugar	Washed raw sugar of large crystals	Halal
Turmeric	A spice and colorant obtained from rhizome or root of *Curcuma longa*	Halal
Tyrosine	An amino acid usually isolated from silk waste	Doubtful
Vanilla extract	A flavorant made from vanilla bean extract	Doubtful
Vanilla flavor, artificial	A flavorant composed of vanillin and ethyl vanillin	Doubtful
Vanillin	A flavorant made from synthetic or artificial vanilla	Doubtful
Vegetable gum	Gums obtained from plant source	Halal
Vegetable oils	Oils obtained from a vegetable source, including soy beans, peanuts, cottonseeds, and plants	Halal

Halal Status for Ingredients (continued)

Ingredient Name	Description	Halal Status
Vinegar	An acidulant and flavorant produced by successive alcoholic and acetous fermentations	Halal
Vitamin K	A fat-soluble vitamin	Halal
Waxy maize starch	The starch protein of waxy corn	Halal
Waxy rice flour	A flour obtained from waxy rice	Halal
Wheat flour	A fine powder obtained by milling wheat	Halal
Wheat germ	The oil-containing portion of the wheat kernel	Halal
Wheat gluten	The water-insoluble complex protein fraction separated from wheat flours	Halal
Wheat starch	A starch obtained from wheat	Halal
Whey	The portion of milk remaining after coagulation and removal of curd	Doubtful
Whey powder	The solid fraction or dry form of whey	Doubtful
Whey protein concentrate	Whey powder where some of the nonprotein has been removed	Doubtful
Whey protein isolate	Proteins isolated from whey	Doubtful
Whole milk solids	The product resulting from the drying or desiccation of milk	Halal
Whole wheat flour	The flour obtained by grinding cleaned wheat	Halal
Wine vinegar	The vinegar made by the alcoholic and acetous fermentation of the juices	Halal
Worcestershire sauce	A sauce consisting of many ingredients	Doubtful
Xanthan gum	A gum obtained by microbial fermentation from *Xanthomonas campestris*	Halal
Xylitol	A sweetener that is as sweet as sucrose	Halal
Yeast extract	A flavor enhancer obtained from the yeast cells of *Saccharomyces cerevisiae*	Halal
Yeast food	A complete food; used in doughs	Doubtful
Yellow prussiate of soda	An anticaking agent and crystallizing agent	Halal

Halal Status for Ingredients (continued)

Ingredient Name	Description	Halal Status
Yogurt	A custard-like or soft gel product	Doubtful
Zein	A corn protein produced from corn gluten meal	Halal
Zinc acetate	A nutrient and dietary supplement	Halal
Zinc carbonate	A nutrient and dietary supplement	Halal
Zinc chloride	A nutrient and dietary supplement	Halal
Zinc gluconate	A source of zinc that functions as a nutrient and dietary supplement	Halal
Zinc oxide	A nutrient and dietary supplement	Halal
Zinc stearate	A nutrient and dietary supplement	Doubtful
Zinc sulfate	A nutrient and dietary supplement; used in frozen substances	Halal

INDEX